DIMENSIONS OF FOOD

FOURTH EDITION

JOIN US ON THE INTERNET
WWW: http://www.thomson.com
EMAIL: findit@kiosk.thomson.com

thomson.com is the on-line portal for the products, services and resources available from International Thomson Publishing (ITP). This Internet kiosk gives users immediate access to more than 34 ITP publishers and over 20,000 products. Through *thomson.com* Internet users can search catalogs, examine subject-specific resource centers and subscribe to electronic discussion lists. You can purchase ITP products from your local bookseller, or directly through *thomson.com*.

Visit Chapman & Hall's Internet Resource Center for information on our new publications, links to useful sites on the World Wide Web and an opportunity to join our e-mail mailing list. Point your browser to: **http://www.chaphall.com** or **http://www.foodsci.com** for Food Science or **http://www.chaphall.com/chaphall/nutrit.html** for Nutrition

---

A service of I(T)P®

DIMENSIONS OF FOOD

FOURTH EDITION

Vickie A. Vaclavik, Ph.D., R.D.
Department of Clinical Nutrition
The University of Texas Southwestern Medical Center at Dallas
Dallas, Texas

Marcia H. Pimentel, M.S.
New York State College of Human Ecology
Cornell University
Ithaca, New York

Marjorie M. Devine, Ph.D.
New York State College of Human Ecology
Cornell University
Ithaca, New York

 CHAPMAN & HALL

 **International Thomson Publishing**
Thomson Science

New York • Albany • Bonn • Boston • Cincinnati • Detroit • London • Madrid • Melbourne
Mexico City • Pacific Grove • Paris • San Francisco • Singapore • Tokyo • Toronto • Washington

Cover design: Andrea Meyer, Emdash Inc.

Copyright © 1998 by Chapman & Hall

Printed in the United States of America

Chapman & Hall
115 Fifth Avenue
New York, NY 10003

Chapman & Hall
2-6 Boundary Row
London SE1 8HN
England

Thomas Nelson Australia
102 Dodds Street
South Melbourne, 3205
Victoria, Australia

Chapman & Hall GmbH
Postfach 100 263
D-69442 Weinheim
Germany

International Thomson Editores
Campos Eliseos 385, Piso 7
Col. Polanco
11560 Mexico D.F
Mexico

International Thomson Publishing–Japan
Hirakawacho Kyowa Building, 3F
1-2-1 Hirakawacho-cho
Chiyoda-ku, 102 Tokyo
Japan

International Thomson Publishing Asia
221 Henderson Road #05-10
Henderson Building
Singapore 0315

All rights reserved. No part of this book covered by the copyright hereon may be reproduced or used in any form or by any means—graphic, electronic, or mechanical, including photocopying, recording, taping, or information storage and retrieval systems—without the written permission of the publisher.

1 2 3 4 5 6 7 8 9 10 XXX 01 00 99 98

Library of Congress Cataloging-in-Publication Data

Vaclavik, Vickie A.
   Dimensions of food / Vickie A. Vaclavik, Marcia H. Pimentel.
Marjorie M. Devine. -- 4th ed.
     p. cm.
   Includes bibliographical references and index.
   ISBN 0-412-14731-9 (alk.paper)
   1. Food--Laboratory manuals. 2. Nutrition--Laboratory manuals.
3. Cookery--Laboratory manuals.   I. Pimentel, Marcia H.
II. Devine, Marjorie M.
TX354.V33   1997
664--dc21   97-27378
                                            CIP

British Library Cataloguing in Publication Data available

To order this or any other Chapman & Hall book, please contact **International Thomson Publishing, 7625 Empire Drive, Florence, KY 41042.** Phone: (606) 525-6600 or 1-800-842-3636. Fax: (606) 525-7778. e-mail: order@chaphall.com.

For a complete listing of Chapman & Hall titles, send your request to **Chapman & Hall, Dept. BC, 115 Fifth Avenue, New York, NY 10003.**

Contents

Preface xi

References xiii

World Wide Web (WWW) Sites xiv

PART I. DIMENSIONS OF FOOD

A. Economic Dimensions — 2

Exercise 1: Factors Influencing Cost of Food — 3
  A. Quality of Product—Comparing Store, National, and Generic Brands — 3
  B. Caloric and Price Differences of Various Product Formulations — 4
  C. The Cost of Convenience Foods—Ready-Made Products, Packaged Mixes — 4
  D. Comparing Price Per Serving of Various Forms of a Food — 5
  E. Cost Comparison of Food, As Purchased (A.P.) and Edible Portion (E.P.) — 7
Exercise 2: Labels as Guides in Food Purchasing — 8
  A. Mandatory Label Information — 8
  B. Labels of Products With "Standards of Identity" — 9
Summary Questions—Economic Dimensions — 10

B. Nutritional Dimensions — 11

Exercise 1: Determining Serving Size — 11
Exercise 2: Factors Affecting Caloric Value of Foods — 12
Exercise 3: Nutrient Contributions of the Food Guide Pyramid — 13
Exercise 4: Evaluation of a Daily Menu — 17
Exercise 5: Labels as Guides to Nutrient Content — 18
  A. Nutritive Value and Cost of Fruit Juice Products — 18
  B. Nutritive Value and Cost of Cereal Products — 19
  C. Carbohydrate Label Information — 20
  D. Health Claims Allowed on Labels — 21
Summary Questions—Nutritional Dimensions — 21

C. Palatability Dimensions — 23

Exercise 1: Identifying Sensory Properties of Food — 23
Exercise 2: Evaluating Sensory Properties in Foods — 26
Exercise 3: Sensory Evaluation Tests — 27
Exercise 4: Evaluating Personal Preferences — 28
Summary Questions—Palatability Dimensions — 28

D. Chemical Dimensions — 29

Exercise 1: Functions of Food Additives — 31
Exercise 2: Relationship of Additive Use to Degree of Processing — 32
Exercise 3: Evaluation of Snack Foods — 33
Exercise 4: Sodium Content of Foods — 34
Summary Questions—Chemical Dimensions — 35

E. Sanitary Dimensions — 37

Exercise 1: Factors Affecting the Microbial Safety of Foods — 38
   A. Sources of Contamination — 39
   B. Conditions Necessary for the Growth of Bacteria — 40
   C. Bacterial Growth Curve — 41
Exercise 2: Temperature Control in Food Handling — 41
   A. Factors Affecting the Rate of Cooling of Large Quantities of Food — 42
   B. Temperatures for Holding and Reheating Foods — 42
   C. Recommended Temperatures for Cooked Food — 43
Exercise 3: Sanitization in the Food Preparation Environment — 43
   A. Use of Approved Chemical Sanitizers — 43
   B. Sanitization by Immersion — 43
Summary Questions—Sanitary Dimensions — 44

F. Food Processing Dimensions — 46

Exercise 1: Processing Temperatures — 47
Exercise 2: Food Processing, Canning — 48
   A. Canning Equipment — 49
   B. Canning Acid and Low-Acid Foods — 49
Questions—Canning — 51
Exercise 3: Food Processing, Freezing — 51
   A. Freezing Equipment — 51
   B. Freezing Fruits and Vegetables — 51
Questions—Freezing — 52
Summary Questions—Dimensions of Food — 53

PART II. FOOD PRINCIPLES

A. Measurements, Use of Ingredients, and Laboratory Techniques — 57

Exercise 1: Demonstration of Measuring and Mixing Techniques — 57
Exercise 2: Measuring Liquids — 58
Exercise 3: Measuring Solids — 58
Exercise 4: Clean-up — 59
Summary Questions—Measurements, Use of Ingredients, and Laboratory Techniques — 59

B. Cereal and Starch — 61

Exercise 1: Separation of Starch Granules — 62
Exercise 2: Properties of Wheat and Cornstarch — 63

Exercise 3: Effect of Sugar and Acid on Gelatinization 64
Exercise 4: Application of Principles to Starch-Thickened Products 65
Exercise 5: Preparing Cereal Products 68
Cereal Recipes 70
Summary Questions—Cereal and Starch 73

C. Fruits and Vegetables
76

Exercise 1: Properties of Parenchyma Cells 77
  A. Components of Parenchyma Cell 77
  B. Recrisping Succulents 78
Exercise 2: Assessing Nutrient Loss in Fruits and Vegetables 79
  A. Effect of Cutting, Chopping, and Soaking on Vitamin C (Ascorbic Acid) 79
  B. Effect of Length of Cooking Time on Vitamin C 80
  C. Effect of Added Alkali on Vitamin C 81
Exercise 3: Fruits 83
  A. Enzymatic Browning 83
  B. Effect of Sugar on Texture and Flavor of Cooked Fresh Fruit 84
  C. Effect of Sugar on Texture and Flavor of Cooked Dried Fruit 85
  D. Factors Affecting Anthocyanin Pigments 86
Exercise 4: Cooking Vegetables 90
  A. Effect of pH on Pigments and Texture 90
  B. Effect of Cooking Procedure on Pigments and Flavors 92
  C. Application of Principles to Cooking a Variety of Vegetables 94
Evaluation of Vegetable Recipes 95
Nutritive Value of Recipes 97
Vegetable Recipes 98
Summary Questions—Fruits and Vegetables 105

D. Meat, Poultry, and Fish
107

Exercise 1: Identification of Basic Meat Cuts 110
Exercise 2: Effect of Dry and Moist Heat on Less Tender (Tough) Cuts of Meat 111
  A. Roasts 111
  B. Meat Patties 113
Exercise 3: Application of Principles to Meat, Poultry, and Fish Cookery 114
Evaluation of Meat, Poultry and Fish Recipes 115
Meat, Poultry and Fish Recipes 117
Summary Questions—Meat, Poultry, and Fish 122

E. Plant Proteins
127

Exercise 1: Pretreatment and Cooking for Legumes/Lentils 128
  A. Pretreatment 128
  B. Cooking Method 129
Exercise 2: Combining Plant Proteins 130
Evaluation of Plant Protein Recipes 131
Plant Protein Recipes 132
Summary Questions—Plant Proteins 137

F. Eggs — 139

Exercise 1: Egg Quality	140
Exercise 2: Coagulation of Egg Protein in Baked and Stirred Custard	141
Exercise 3: Egg White Foams	144
Exercise 4: Effect of Added Substances on Egg White Foam	145
Exercise 5: Effect of Cooking Intensity on Egg Protein	147
Exercise 6: Characteristics of Cooked Modified Egg Mixtures	149
Exercise 7: Combining Starch and Eggs as Thickeners in One Product (Soufflé)	150
Soufflé Recipes	151
Summary Questions—Eggs and Egg Products	154

G. Milk and Milk Products — 157

Exercise 1: Comparison of Milk and Non-Dairy Products	158
Exercise 2: Coagulation of Milk Protein	159
A. Addition of Acid	159
B. Acid Produced by Bacteria (Yogurt)	159
C. Enzyme Action (Rennin)	160
Exercise 3: Combining Acid Foods with Milk	160
Exercise 4: Comparison of Cheese Products	161
Exercise 5: Effect of Heat on Natural and Processed Cheese	164
Summary Questions—Milk and Milk Products	165

H. Batters and Doughs — 168

Exercise 1: Measurement of Flour	169
Exercise 2: Structural Properties of Wheat Flour	169
Exercise 3: Chemical Leavening Agents	171
A. Ingredients of Baking Powders	171
B. Comparison of Speed of Reaction	171
Exercise 4: Factors Affecting the Leavening Power of Yeast	172
Questions—Leavening Agents	172
Exercise 5: Drop Batter, Muffins	173
A. Effect of Manipulation	174
B. Effect of Different Grains	175
Questions—Muffins	176
Exercise 6: Soft Dough, Biscuits	177
A. Effect of Manipulation	178
B. Substituting Soda-Acid for Baking Powder	179
Questions—Biscuits	179
Exercise 7: Pancakes, Popovers, Cream Puffs	180
A. Effect of Manipulation on Gluten Development in Pancakes	180
B. Effect of Manipulation on Gluten Development in Popovers	181
C. Cream Puffs	182
Questions—Pancakes, Popovers, Cream Puffs	183
Exercise 8: Stiff Dough-Yeast Bread/Rolls	184
Evaluation of Yeast Rolls	185
Questions—Yeast Rolls	186
Exercise 9: Shortened Cakes	187
Questions—Cakes	189

Exercise 10: Stiff Dough-Pastry — 190
   A. Effect of Different Fat Plasticity on Palatability of Pastry — 190
      Evaluation of Pastry — 191
   B. Effect of Different Fillings on Palatability of Bottom Crust — 192
Questions—Pastry — 195
Summary Questions—Batters and Doughs — 197

I. Fats and Oils

Exercise 1: Separation and Ratio of Oil and Acid; Emulsifiers — 201
Exercise 2: Application of Principles to Salad Dressings — 202
Salad Dressing Recipes — 202
Exercise 3: Fat-Free, Fat-Reduced, and Fat Replaced Products — 203
   A. Cost and Palatability of Fat-Free, Fat-Reduced and Fat Replaced Products — 205
   B. Fat Substitute Labeling — 205
Exercise 4: Comparison of Dietary Fats — 206
Summary Questions—Fats and Oils — 207

J. Sugars, Sweeteners

Exercise 1: Methods of Initiating Crystallization — 210
Exercise 2: The Relationship of Sugar Concentration to Boiling Point — 211
Exercise 3: Effect of Temperature and Agitation on Crystal Size — 212
Exercise 4: Effect of Interfering Agents on Sugar Structure — 213
Ecercise 5: Sugar Substitutes, High-Intensity Sweeteners — 214
Summary Questions—Sugars, Sweeteners — 216

PART III. HEATING FOODS BY MICROWAVE

Microwave Cooking

Exercise 1: Effect of Cooking Procedure on Pigments and Flavors — 220
Exercise 2: Fruits — 221
Exercise 3: Vegetables — 222
Exercise 4: Starch Products — 222
   A. Pasta, Rice, and Cereals — 225
   B. Flour and Cornstarch as Thickeners — 225
Exercise 5: Eggs — 226
Exercise 6: Meat, Poultry, and Fish — 227
Exercise 7: Batters and Doughs — 228
Exercise 8: Reheating Baked Products — 229
Exercise 9: Defrosting — 230
Summary Questions—Microwave Cooking — 231

PART IV. MEAL MANAGEMENT

Meal Management

Exercise 1: Analyzing Menus for Palatability Qualities — 235
Exercise 2: Economic Considerations in Menu Planning — 236

Exercise 3: Low Calorie Modifications ... 237
Exercise 4: Meal Planning ... 238
Exercise 5: Meal Preparation ... 244
Summary Questions—Meal Management ... 245

Appendices ... 248

A. Legislation Governing the Food Supply ... 248
B. Food Guides and Dietary Guidelines ... 251
C. Some Food Equivalents ... 255
D. Average Serving or Portion of Foods ... 256
E. Food Additives ... 257
F. pH of Some Common Foods ... 261
G-I. Major Bacterial Foodborne Illnesses ... 262
G-II. Meat and Egg Cooking Regulations ... 263
H. Heat Transfer ... 264
I. Symbols for Measurements and Weights ... 267
J. Notes on Test for Presence of Ascorbic Acid ... 268
K. Cooking Terms ... 269
L. Buying Guide ... 271
M. Fruit and Vegetable Availability ... 274
N. Spice and Herb Chart ... 276
O. Plant Proteins ... 277

Preface

This fourth edition of *Dimensions of Food* reflects the many advances occurring in the food and nutrition fields.

Students of food science, nutrition and dietetics, as well as teachers, and others who are preparing for food-related careers in the twenty-first century understand the important connections between diet and optimum health. They recognize that nutrition, food safety, and the economics of the marketplace are important issues in their personal and professional lives. For some, these interests are coupled with a desire to express a personal value system through the selection and preparation of food. Others wish to take advantage of the technological avalanche of new products to match the fast-paced lifestyles. Simultaneously, many students are investigating and preparing for a food-related career. Such diverse objectives require an integrated approach in their introductory study of food. Our aim is to provide a variety of stimulating exercises and laboratory discussions through which a student can explore and better understand the multidimensional nature of food decisions and preparations.

In-depth study of food economics and labeling as well as assessment of nutrient quality of diets based on the use of the Food Guide Pyramid is presented in Part I. Food safety and sanitary quality, the role of food additives, and principles of preservation are included. Assessment of the palatability characteristics of food, important in personal food choices, is stressed.

Demonstrations and student experimentations of Part II provide the experiences basic to understanding the functional and structural properties of the components of all foods. In line with current interests in nutrition, exercises emphasize the preparation of fruits and vegetables, varieties of grains, as well as plant proteins. Batters and doughs continue to be a major area of study. Other chapters are devoted to dairy products, meats, and fish; the role of fats in food preparation; the principles of crystallization; and artificial sweeteners. Microwave cookery, Part III, can be planned as a separate unit or easily integrated into the exercises of Part II.

Throughout the manual careful attention is given to the preserving of major nutrients and palatability quality. Recipes have been revised to reduce total fat, saturated fat, sodium, and cholesterol, yet retain flavor and appeal. Also, recipe selection has been expanded to more completely represent cultural and geographic diversity.

The creative use of all dimensions of food is applied in the meal-planning and preparation exercises in Part IV. Preliminary exercises have been included to help students understand key principles. Varying cost level and nutritional problems are suggested. These exercises may be summarizing experiences or used at various points during the course. Even if time does not permit inclusion of this section in an introductory course, Part IV serves as a reference for those students who are unable to take the advanced course.

Throughout the manual, learning experiences are sequences to move from basic demonstrations of key principles to their application. Study questions and problems are designed to help students clarify and organize facts into working principles.

The updated appendices supplement current textbooks and provide additional background information for the exercises. Nutrition information has been expanded to include a discussion of dietary guidelines and the Food Guide Pyramid. Current information about food legislation and mandated food labeling is featured. In view of the latest scientific data available, the sections dealing with foodborne illnesses have been expanded.

Our aim is to provide a variety of experiences from which an instructor may choose those most helpful to the student. Activities may be carried out in the laboratory, demonstrated, or assigned as projects to be completed outside of the classroom. Once completed the laboratory manual serves as a valuable personal and reference.

Many people have contributed to the evolution of this manual. Special acknowledgments are extended to our colleagues for their careful review of parts of the manuscript and to our many teaching assistants for their helpful contributions. To our students, whose curiosity and penetrating questions continue to make teaching the dimensions of food a joyful challenge, we say, thank you!

V.A. Vaclavik (Vä klä′ vĭk)
M.H. Pimentel
M.M. Devine

REFERENCES

Agricultural Handbook No. 8 (all volumes). Revised, 1976–1990. *Composition of Food—Raw, Processed, Prepared.* Washington, DC: Agricultural Research Center, USDA, 1990.

Educational Foundation of the National Restaurant Association. *Applied Foodservice Sanitation*, 4th ed. New York: John Wiley & Sons, 1992.

Food Guide Pyramid. Washington, DC: U.S. Department of Agriculture.

Handbook of Food Preparation. 9th ed. Kendall-Hunt Publishing Company, Dubuque, IA: American Home Economics Association, 1991.

Molt M. *Food for Fifty*, 10th ed. Englewood Cliffs, NJ: Prentice-Hall, 1997.

Pennington JAT. *Bowes and Church's Food Values of Portions Commonly Used*, 16th ed. Philadelphia: JB Lippincott, 1994.

Recommended Dietary Allowances, 10th ed. Washington, DC: Food and Nutrition Board, National Academy of Sciences. 1989.

U.S. Department of Agriculture and US Department of Health and Human Services. *Nutrition and Your Health: Dietary Guidelines for Americans*, 4th ed. Washington, DC: U.S. Government Printing Office, 1995.

Vaclavik VA. *Essentials of Food Science.* New York: Chapman & Hall, 1997.

World Wide Web (WWW) Sites

FDA - http://www.fda.gov

FDA Federal Regulations - http://www.access.gpo.gov/nara/cfr

USDA - http://www.usda.gov

USDA database - http://www.nal.usda.gov.fnic/foodcomp

The American Dietetic Association - http://www.eatright.org

The Institute of Food Technologists - http://www.ift.org

The National Restaurant Association - http://www.restaurant.org

Part I

Dimensions of Food

To maximize health and pleasure, food must be nutritious, safe to eat, personally satisfying, and obtainable within the resources that each of us chooses to expend. Realizing these goals in a complex marketplace is a confrontation—a confrontation of values, resources, choices, and conflicting information.

Part I of this manual helps clarify and examine in some depth, the multidimensional nature of food decisions. Exercises are concerned with the economic, palatability, legislative, sanitary, chemical, nutritional, and processing dimensions of our food supply. Questions and exercises are provided to help organize key principles into yardsticks by which the relative values of food can be appraised and food choices tailored to personal resources and beliefs.

A. Economic Dimensions

OBJECTIVES

To recognize factors that influence cost of food items
To calculate and compare cost per unit of various food items
To identify types of information available to the consumer in the marketplace
To delineate uses of different food qualities for specific purposes
To interpret, evaluate and use food label information as a buying guide
To distinguish and enumerate the many considerations involved in choosing a "best food buy" for an individual or family

REFERENCES

Appendix A

ASSIGNED READINGS

TERMS

Standards of Identity	House brand	A.P.
Food, Drug and Cosmetic Act	Generic brand	E.P.
Miller Pesticide Amendment	NAS	Unit Pricing
Food Additive Amendment	NRC	Gross weight
Delaney Clause	USDA	Net weight
Fair Packaging Act	RDA	FDA
Labeling regulation	Daily Value	Nutrition Facts

In the following sections, exercises are presented to illustrate some factors that affect the economy of food and food decisions.

EXERCISE 1: FACTORS INFLUENCING COST OF FOOD

PROCEDURE

1. Complete the tables by matching the samples of products displayed with the price/item *or* by visiting a grocery market.
2. Compare information and explanations with classmates.

A. QUALITY OF PRODUCT—COMPARING STORE, NATIONAL, AND GENERIC BRANDS

Brand	Price/ Product	Price/ Net oz (g)	Description	Uses of Product
National brand #1				
National brand #2				
Store brand				
Generic brand				

1. Why is there a difference in price/can of the same product?

2. What is the difference between a store brand and generic brand?

3. What do the different labels state as to type of pack and net contents?

4. Which do you consider the "best buy"? What criteria are used by consumers in selecting canned products?

4 DIMENSIONS OF FOOD

B. CALORIC AND PRICE DIFFERENCES OF VARIOUS PRODUCT FORMULATIONS

Evaluate the relative caloric and price differences among the regular and "special" diet products, e.g., canned fruit, salad dressings, soft drinks, etc.

Brand and Product	Calories/Serving	Cost/Serving
1 a. (reg.)		
b. (diet)		
2 a.		
b.		
3 a.		
b.		

1. Does the caloric value justify the price of any of these products? Explain.

2. How does caloric information and unit pricing assist in the consumer's food purchasing selection?

C. THE COST OF CONVENIENCE FOODS—READY-MADE PRODUCTS, PACKAGED MIXES

Examine several ready-made and packaged mixes of convenience foods. Compare cost per serving of convenience foods to the same product prepared from scratch, e.g., baked items, puddings, etc.

Type of Food	Brand and Product	Cost/Serving
1. "Scratch" Form		
Ready-Made Form		
Packaged "Mix"		
2. "Scratch" Form		
Ready-Made Form		
Packaged "Mix"		

1. What is the relationship between convenience and price?

2. What factors may influence a consumer's choice of product form?

D. COMPARING PRICE PER SERVING OF VARIOUS FORMS OF A FOOD

1. Potatoes

Compare several potato products. Calculate price/serving[1] and suggest possible uses.

Form of Potato	Price/Unit	Price/Serving	Possible Uses of Product
Idaho baker			
Potatoes (regular)			
Canned potatoes			
Dried potato flakes			
Frozen french fries			
Frozen hash browns			

[1] See Appendix C.

1. What factors influence the form of potatoes that will be chosen for dinner by an individual?

2. How will the season of the year affect the above price comparison?

3. How does the geographic location of a consumer affect the price that he/she pays for food?

2. Milk

Compare several milk products and complete the following chart[1]:

Form of Milk	Unit	Cost/ Unit	Cost/ 8 oz (240 ml)	Uses of Product
Nonfat dry milk (national brand)	Bulk			
Nonfat dry milk (store brand)	Bulk			
Nonfat dry milk (generic brand)	Bulk			
Nonfat dry milk	qt (L) pkg			
Evaporated milk	12 oz (360 ml)			
Evaporated nonfat	12 oz (360 ml)			
Fresh, whole	1 qt (1 L)			
Fresh, 2%	1 qt (1 L)			
Fresh, 1%	1 qt (1 L)			
Fresh, ½%	1 qt (1 L)			
Fresh, nonfat	1 qt (1 L)			

[1]For laboratory purposes, use quart or liter as market unit.

1. How does the form and brand of milk affect price?

2. State how "unit pricing" assists consumers in determining the best buy for the money spent.

3. Discuss how the various forms of milk could be used in order to take advantage of price differences.

E. COST COMPARISON OF FOOD, AS PURCHASED (A.P.) AND EDIBLE PORTION (E.P.)

Compare the cost per pound (454 g) of food, as purchased (A.P.) to cost per pound (454 g), edible portion (E.P.).

Food	Cost/lb (g) A.P.	Percent Waste[a]	Weight E.P.	Cost/lb (g) E.P.
Apple		8		
Banana		32		
Broccoli		22		
Carrot		18–22		
Orange		25–32		
Fresh peas (in pod)		62		
Chicken, fryer		32		
Haddock, fillet		0		

[a]*Source*: Watt BK, Merrill AL. *Composition of Foods: Raw, Processed, Prepared.* Washington, DC: Agricultural Handbook No. 8, Consumer and Food Economics Research Division, USDA, 1963.

1. What factors influence the percentage of waste in a food?

2. How is information about percent waste of value to the consumer?

8 DIMENSIONS OF FOOD

EXERCISE 2: LABELS AS GUIDES IN FOOD PURCHASING

PROCEDURE

1. Complete the tables by viewing the samples of products displayed or by visiting a grocery market.
2. Share information and answers with classmates.

A. MANDATORY LABEL INFORMATION

Compare labels on containers on several kinds of food items.

Brand and Product	Net Weight	Manufacturer Name & Address	Ingredients Listed	Other Label Information

1. What information must be on all food labels?

2. What does the listing of ingredients tell about relative amounts of ingredients in a product?

3. What government regulation specifies criteria for labeling?

4. Which federal agency is charged with enforcing the above regulation?

5. What label format or information not currently provided, would be useful to consumers?

B. Labels of Products With "Standards of Identity"

Study the labels of several bread, mayonnaise, ketchup, pasta, or jelly products. Note the listing of ingredients.

Brand and Product	List of Ingredients

1. What is the purpose of a standard of identity?

2. Are some products allowed to omit a complete listing of ingredients?

3. Should manufacturers of all food products be required to provide a complete list of ingredients?

The New Food Label at a Glance

Claims: While descriptive terms like "low," "good source," and "free" have long been used on food labels, their meaning — and their usefulness in helping consumers plan a healthy diet — have been murky. Now FDA has set specific definitions for these terms, assuring shoppers that they can believe what they read on the package.

Ingredients still will be listed in descending order by weight, and now the list will be required on almost all foods, even standardized ones like mayonnaise and bread.

Health Claims: For the first time, food labels will be allowed to carry information about the link between certain nutrients and specific diseases. For such a "health claim" to be made on a package, FDA must first determine that the diet-disease link is supported by scientific evidence.

"While many factors affect heart disease, diets low in saturated fat and cholesterol may reduce the risk of this disease."

Health claim message referred to on the front panel is shown here.

Source: Food and Drug Administration, 1994

SUMMARY QUESTIONS—ECONOMIC DIMENSIONS

1. Summarize the diverse factors that influence prices of food.

2. Based on comparisons of various food products, list major factors a consumer might consider in selecting good money buys.

3. In practice, what factors, other than price, do consumers take into account when selecting food in the marketplace?

B. Nutritional Dimensions

OBJECTIVES

To identify standard serving size of selected foods
To identify factors influencing caloric value of foods
To compute nutritive values of selected foods
To generalize the major nutrient contributions of groups in the Food Guide Pyramid
To recognize the advantages and disadvantages of food guides
To use the Food Guide Pyramid and Dietary Guidelines to evaluate a dietary plan

REFERENCES

Appendices B, C, D

Pennington, JAT, *Bowes and Church's Food Values of Portions Commonly Used*, 16th ed. Philadelphia: JB Lippencott, 1994.

ASSIGNED READINGS

TERMS

Food Guides
Food Guide Pyramid
Dietary Guidelines
RDA

Standard serving size
NRC
NAS

EXERCISE 1: DETERMINING SERVING SIZE

PROCEDURE

1. Observe a demonstration or calculate the standard serving size of several foods or food models.
2. Record weight/measure in the table.
3. Display food or food models.
4. Compare your personal conception of a serving with the standard serving

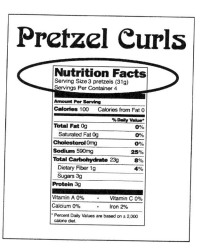

Courtesy: FDA

Food	Standard Serving Size weight/measure	Comments
Bread		
Crackers		
Pasta		
Peanut butter		
Meat, poultry, fish		
Cooked legumes		
Cooked vegetable		
Canned fruit		
Whole fruit		
Other		

EXERCISE 2: FACTORS AFFECTING CALORIC VALUE OF FOODS

PROCEDURE

1. Observe a demonstration or calculate 100-calorie portions of apple products.
2. Display and compare size of portions. Record observations.

	Raw Apple	Applesauce	Apple Pie
100-kcal portion size			

What are some of the factors that affect the caloric value of these apple products?

EXERCISE 3: NUTRIENT CONTRIBUTIONS OF THE FOOD GUIDE PYRAMID

Procedure

1. Determine the standard serving size of foods in the group of foods assigned.

A	B	C
skim milk	whole milk	apple
egg	beets	cheddar cheese
frozen peas	pork chop	hamburger
highly fortified cereal	canned peas	macaroni, unenriched
lima beans	corn flakes	potato, white
peanut butter	puffed rice cereal	sweet potato/yams
lettuce	black-eyed peas	bread, unenriched
canned pineapple	process cheese	tortilla

D	E	F
canned peaches	celery	cottage cheese
yogurt	margarine	biscuit
tomato	spinach/kale	fresh green beans
hot dog	orange	bologna
collard greens	tuna fish	fish fillet
kidney beans	tomato juice	cabbage
rice/grits	bread, enriched	pancakes
carrots, raw	grapes	frosted cereal

2. Calculate and record the nutritive value of servings under the appropriate headings on chart.
3. Display servings of foods or food models, with a chart of completed nutritive value. Arrange foods on display table according to the Food Guide Pyramid.

Nutritive Contributions of the Food Guide Pyramid

Food Group	Wt/Meas per serving	Energy (kcal)	Protein (g)	Fat (g)	Carbohydrates (g)	Vitamin A (IU or R.E.)	Vitamin C (mg)	Calcium (mg)	Iron (mg)
Bread, cereal, rice, and pasta									
Vegetables									

(Continued)

Nutritive Contributions of the Food Guide Pyramid (Cont'd)

Food Group	Wt/Meas per serving	Energy (kcal)	Protein (g)	Fat (g)	Carbohydrates (g)	Vitamin A (IU or R.E.)	Vitamin C (mg)	Calcium (mg)	Iron (mg)
Fruit									
Milk, yogurt, and cheese									

(Continued)

Nutritive Contributions of the Food Guide Pyramid (Cont'd)

Food Group	Wt/Meas per serving	Energy (kcal)	Protein (g)	Fat (g)	Carbohydrates (g)	Vitamin A (IU or R.E.)	Vitamin C (mg)	Calcium (mg)	Iron (mg)
Meat, poultry, fish, dry beans, eggs, and nuts									
Fats, oils, and sweets									

EXERCISE 4: EVALUATION OF A DAILY MENU

PROCEDURE

Using the charts below, evaluate the following by the Food Guide Pyramid.

Breakfast	Lunch	Dinner

Orange Juice (½ c [120ml])
Cheese Omelet (2 eggs, ½ oz [14 g] cheese)
Whole wheat toast (2)
   Buttered (½ t [2.5 ml])
Strawberry Jam (2 T [30 ml])
	Navy Bean Soup
	   (¼ c [60ml] beans)
	Saltines (4)
	Spinach Salad (1 c [240 ml])
	Yogurt Dressing (¼ c [60 ml])
	Fig Bar Cookie (2)
		Chicken Tetrazzini
		   (2 oz [57 g] chicken,
		   ½ oz [14 g] cheese,
		   ½ c [120 ml] spaghetti)
		Green Beans (¾ c [180ml]),
		   Buttered (½ t [2.5ml])
		Carrot Sticks (½ c [120 ml])
		Strawberry Gelatin (½ c [120ml])
		Milk, Whole (8 oz [240 ml])

Bread, Cereal, Rice, and Pasta	Vegetable	Fruit	Milk, Cheese, and Yogurt

Meat, Poultry, Fish, Dry Beans, Eggs, and Nuts	Fats, Oils, and Sweets	Foods Difficult to Classify

EXERCISE 5: LABELS AS GUIDES TO NUTRIENT CONTENT

PROCEDURE

1. Complete the tables by viewing samples of products displayed or by visiting a grocery market.
2. Share information and answers with classmates

A. NUTRITIVE VALUE AND COST OF FRUIT JUICE PRODUCTS

Study the labels of several fruit juice products (e.g., punch, ade, frozen, formulated) and compare costs and vitamin C content.

Brand and Product	Cost/cup (240 ml)	Vitamin C/cup (240 ml)	Fruit Juice (%)

1. Considering the vitamin C content, which is the best buy? Why?

2. Considering the percentage natural fruit juice content, which is the best buy?

3. What other criteria for judgment is important to you as a consumer of a juice product? Explain.

B. NUTRITIVE VALUE AND COST OF CEREAL PRODUCTS

Study the labels of several dry cereal products currently on the market. Complete the following chart, selecting two cereals for comparison.

Brand of Cereal		
	1.	2.
Serving Size		
Servings Per Container		
Amount Per Serving		
Calories Calories from Fat		
Total Fat Saturated Fat		
Cholesterol		
Sodium		
Total Carbohydrate		
Dietary Fiber		
Sugars		
Other Carbohydrates		
Protein		

1. Were all the cereal labels equally helpful in determining the nutritive value? Explain.

2. Based on the label information, which cereal is the best "nutritive" buy? Which is the "best buy" for you? Why?

C. CARBOHYDRATE LABEL INFORMATION

PROCEDURE

Study the illustration of carbohydrate information that appears on cereal product labels:

	1 oz (28 g) Serving
Starch and related carbohydrates	12 g
Sucrose and other sugars	5 g
Dietary fiber	3 g
Total carbohydrates	20 g

1. Specify the difference between the two carbohydrates, sugar and starch.

2. Concerning the 5 g of sugar in this 1-oz serving, what is the measure equivalent?

3. Define "dietary fiber." How does it differ from "crude fiber"?

D. HEALTH CLAIMS ALLOWED ON LABELS

Specify the allowed nutrient disease relationship claims and rules for their use, for each of the following:

Fat and cancer

Fruits and vegetables and cancer

Fiber-containing grain products, fruits, and vegetables and cancer

Saturated fat and cholesterol and coronary heart disease (CHD)

Fruits, vegetables, and grain products that contain fiber and risk of CHD

Sodium and hypertension (high blood pressure)

Calcium and osteoporosis

Folate and neural tube defect

Sugar alcohols and dental caries

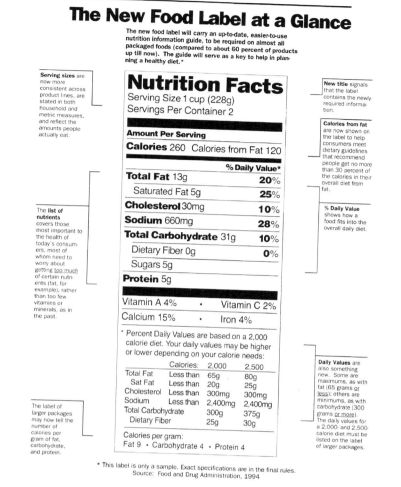

SUMMARY QUESTIONS—NUTRITIONAL DIMENSIONS

1. Provide examples of how the caloric value of some vegetables and meats can be increased by methods of preparation.

2. Explain what is meant by "hidden calories." List several examples to illustrate.

3. Consider your own personal food preferences and habits. What individual foods in each group of the Food Guide Pyramid act as personal safeguard foods for obtaining a regular supply of the major nutrients?

4. Of current concern are the amounts of fat and cholesterol in the American diet. List several foods that supply significant amounts of these items. List suggestions for improving the diet.

5. Consider the foods that comprise the Food Guide Pyramid. What generalizations can be made as to the major nutritive contributions of each group?

Group	Protein	Carbohydrate	Fat	Vitamin A	Vitamin C	Iron	Calcium
Bread, cereal, rice, and pasta							
Vegetables							
Fruit							
Milk, cheese, and yogurt							
Meat, poultry, fish, dry beans, eggs, and nuts							

6. As the director of a Head Start school, you are concerned about the feeding program available to your children, especially breakfast. Because your cook is not on duty until 9 a.m., some provision must be made to serve a nutritious breakfast, yet one which does not involve cooking. You are considering a new product called "Magic Muffin" which tastes like a frosted cupcake, but is highly fortified. Briefly discuss some of the aspects you will need to consider before you make a final decision.

7. a. List two advantages to the consumer of the Nutrition Facts labels on food products.

   b. What additional nutrition information would you like to see on labels of food products?

C. Palatability Dimensions

OBJECTIVES

To identify major sensory properties of food
To describe sensory characteristics responsible for perception of flavor
To evaluate a product in various forms as to sensory properties and personal preference
To identify various sensory tests used for food acceptability evaluation

REFERENCES

ASSIGNED READINGS

TERMS

Appearance   Aroma   Off-flavor   Temperature
Texture   Tenderness   Mouthfeel   Subjective Testing
Flavor   Moistness   Consistency   Sensory Testing

EXERCISE 1: IDENTIFYING SENSORY PROPERTIES OF FOOD

PROCEDURE

1. Sample foods of each category listed on the chart.
2. Categorize the predominant sensory property(ies).
3. Describe sensory properties as fully as possible. Compare and contrast samples in each section of the chart. Discuss observations with classmates.

> Principal Sensory Properties of Food
> 1. Appearance
> 2. Flavor—taste and odor
> 3. "Mouthfeel" texture
> 4. Odor or aroma
> 5. Temperature

24 DIMENSIONS OF FOOD

Food	Predominant Sensory Property(ies)	Description
A. 1. Standard—1 cup water (240 ml)		
2. 1 c water + 1 t sugar (5 ml)		
3. 1 c water + 1 T sugar (15 ml)		
4. 1 c water + 2 t lemon juice (10 ml)		
5. 1 c water + 2 t lemon juice (10 ml) 2 t sugar (10 ml)		
6. 1 c water + ½ t salt (2.5 ml)		
7. 1 c water + ½ t salt (2.5 ml) ½ t sugar (2.5 ml)		
B. 1. Raw onion		
2. Raw apple		
3. Raw potato		
C. 1. Cold ice cream		
2. Melted ice cream		
3. Gelatin (solid)		
4. Gelatin (liquid soft)		
D. 1. Mineral oil		
2. Crackers		
3. Celery		
4. Angel food cake		
5. Pickles		
6. Mints		

(Continued)

Food	Predominant Sensory Property(ies)	Description
7. White bread		
E. Sample plain tomato juice. Add different seasonings singly and taste. Rinse mouth with water between samples. Compare and contrast the effect of different seasonings. Examples: sugar, salt, lemon juice, basil, Worcestershire sauce, tarragon, oregano		
F. Other		
1.		
2.		
3.		

1. How does temperature affect perception of flavor?

2. How do basic tastes differ when used together?

3. For what foods is odor or aroma the predominant sensation?

4. Explain how a "standard", as used in Part A, is helpful in evaluating sensory properties.

EXERCISE 2: EVALUATING SENSORY PROPERTIES IN FOODS

PROCEDURE

1. Compare sensory properties of a single product that has been processed in different ways.
   Examples: Macaroni and cheese: "scratch," packaged, canned, frozen
   Vanilla pudding: regular, instant, canned
   Corn: fresh, frozen, canned, creamed
   Baked item: "scratch," mix, prepared
2. Describe the blend of sensory properties characteristic of any product that was sampled.
3. State which product was preferred. Explain your choice based on personal preference and sensory properties.

	Sensory Properties	Preferences and Reasons
1. a.		
b.		
c.		
d.		
2. a.		
b.		
c.		
d.		

Conclusions:

EXERCISE 3: SENSORY EVALUATION TESTS

Procedure

1. Design an experiment using the following tests for a product and/or a new ingredient in a product[1]:
   a. Paired Comparison Test ("likability")
      A comparison of two food samples to evaluate specific attributes (e.g., color, flavor, texture).

   b. Triangle Test
      A comparison of three samples, including one different item out of three, to determine whether the different attribute can be detected.

   c. Ranking
      A comparison of several food samples by ranking them according to preference.

2. Taste various food items evaluating products using the assigned tests.

3. Record observations and critique of the test used.

Food Item	Test Used	Critique of Test Used
1.		
2.		
3.		
4.		

[1] Suggested products: Beverage sweetened with aspartame/sugar
Regular/lite cheese
Regular/low-salt or no-salt crackers

28 DIMENSIONS OF FOOD

EXERCISE 4: EVALUATING PERSONAL PREFERENCES

PROCEDURE

1. Survey several classmates regarding their food preferences, using the following questionnaire format.
2. List four favorite foods, then check categories that are associated with those foods.

Favorite Foods	Family	Peers	Comfort	Celebration	Nutrition	Other
1.						
2.						
3.						
4.						

3. List four disliked foods, then check the palatability characteristics associated with those foods.

Disliked Foods	Appearance	Flavor	Texture	Other
1.				
2.				
3.				
4.				

SUMMARY QUESTIONS—PALATABILITY DIMENSIONS

1. Discuss how perception of "eating quality" may be influenced by past experiences.

2. Analyze your favorite food, identifying sensory properties and personal factors that make it your favorite.

3. Based on your survey, what are the major palatability characteristics that influence the acceptability of foods?

4. With another individual, select a food that you both like. Together, write a description of the desired properties or the standards you would expect. Do you both agree on all points? Check your description against standard descriptions (product standards in this manual or textbooks).

D. Chemical Dimensions[1]

OBJECTIVES

To recognize the general role of food additives in specific foods
To identify from food labels the ingredients classified as food additives
To understand the laws regulating the use of additives
To relate the extent of processing and use of additives
To evaluate and interrelate the nutritional value, cost and use of additives

REFERENCES

Appendix E

Pennington JAT. *Bowes and Church's Food Values of Portions Commonly Used*, 16th ed. Philadelphia: JB Lippincott, 1994.

ASSIGNED READINGS

TERMS

Enrichment	Delaney Clause	Risk	Maturing agent
Fortification	GRAS	Antioxidant	Humectant
Food additive	Toxicity	Emulsifier	Sequestrant
Food Additive Amendment	Hazard	Stabilizer	Synergist

---

[1] All foods are composed of chemicals. The focus here is on nutritive and non-nutritive chemicals added to foods for specific purposes.

GRAS INGREDIENTS IN COMMON FOODS

Breads	Soft Drinks	Cheeses	Cake Mixes	Canned Fruits/ Vegetables
Preservatives Sequestrants Surfactants Bleaching agents Nutrients	Preservatives Antioxidants Sequestrants Thickeners Acids Coloring agents Non-nutritive sweeteners Nutrients Flavoring agents	Preservatives Sequestrants Thickeners Acids Coloring agents	Antioxidants Sequestrants Surfactants Thickeners Bleaching agents Acids Coloring agents Non-nutritive sweeteners Flavoring agents	Antioxidants Thickeners Alkalis Coloring agents Non-nutritive sweeteners

The Sciences, Vol 14, No. 5, July/August, 1974. © The New York Academy of Sciences. Reprinted with permission.

EXERCISE 1: FUNCTIONS OF FOOD ADDITIVES

PROCEDURE

1. List several major functions of food additives.
2. Study the labels of the foods on display in the laboratory or visit a supermarket.
3. Identify additives that illustrate the functions listed.
4. Note also the products in which the additives are found.
5. Record all information on the following chart.

Specific Additive	Additive Function	Products Containing Additive

EXERCISE 2: RELATIONSHIP OF ADDITIVE USE TO DEGREE OF PROCESSING

PROCEDURE

1. Compare food products[1] prepared from "scratch," prepared mix, and purchased ready to use.
2. Complete the following table after analyzing the ingredients used in the preparation of the various products.

Ingredients		
"Scratch"	Packaged Mix	Ready to Use

[1]Suggested products: main dish item, bakery item, pudding, and gravy.

1. Summarize the relationship between degree of processing and additive use.

2. Identify reasons why one form of the product may be more beneficial to a consumer than the alternatives.

EXERCISE 3: EVALUATION OF SNACK FOODS

PROCEDURE

1. Calculate from appropriate tables and complete the chart with weight and measure of 100-calorie portions of the assigned snack foods[1].
2. If the snack foods are available, place on display and label with weight and measure. Note additives.
3. Star those snacks that are high in simple sugars; double-star those high in fat. Check those snacks that have a high sodium content.

Food Item	Weight and Measure/ 100-kcal portion	Nutrients (Major)	Additives

Conclusions:

---

[1] Suggestions: potato chips; pretzels; air-popped, buttered popcorn; sour cream dip; peanuts; pizza; chocolate bar; bagel, cream cheese; apple; orange; carbonated drink; chocolate chip cookie; ice cream, and so forth.

EXERCISE 4: SODIUM CONTENT OF FOODS

PROCEDURE

Using appropriate references, complete the following chart on the sodium content of common foods.

Food Item	Measure	Sodium Content (mg)	No/Low-Sodium Food (mg)
Tomato Fresh	1 medium		
Canned	1 c (240 ml)		
Soup	1 c (240 ml)		
Ketchup	1 T (15 ml)		
Potato Baked	1 medium		
Chips	1 oz (28 g)		
Dairy Milk	1 c (240 ml)		
Cheddar cheese	1 oz (28 g)		
Meat Beef, ground	3 oz (85 g)		
Chicken	3 oz (85 g)		
Hot dog	1		
Lunch meat	1 slice		
Bread	1 slice		
Flour	½ c (120 ml)		
Crackers	1 serving		
Cucumber	½ large		
Dill pickle	½ large		
Soy sauce	1 T (15 ml)		

1. What is the daily milligram intake of sodium suggested by dietary goals?

2. How many milligrams of sodium are there in 1 teaspoon of salt?

3. List several low-sodium and no-sodium processed foods that are now available in the marketplace.

SUMMARY QUESTIONS—CHEMICAL DIMENSIONS

1. Are all ingredients listed on a label classified as chemicals?

2. Are all ingredients listed on a label classified as additives?

3. What government regulation governs the listing of additives?

4. Provide examples of major additive functions and illustrate each with specific examples.

HIDDEN SALT
Source: Division of Nutritional Sciences, New York State College of Human Ecology at Cornell.

5. Under what circumstances are additives removed from the GRAS list?

6. Based on laboratory evaluation of snack food, what are the major nutrients in most snack food? What are the inadequacies of snack food?

7. Based on a comparison of ingredient lists on labels, what general observations can be made as to factors that influence the number of additives used in a food?

8. Discuss the choices that a consumer has regarding the number of food additives he/she consumes.

9. Discuss the usefulness of label information to consumers who have food intolerances or food allergies.

E. Sanitary Dimensions

OBJECTIVES

To identify factors influencing growth of microorganisms
To relate the environmental needs of bacteria
To describe the phases in the growth of bacteria
To relate key principles for evaluating sanitary quality of food
To recognize the interrelationship of the nature of food, sources of contamination, and time-temperature history of food to its sanitary quality
To describe proper sanitation of food preparation equipment

REFERENCES

Appendices F, G–I, II Local Environmental Health Code

ASSIGNED READINGS

TERMS

Spoiled
Contaminated
Foodborne illness
Intoxication
Infection
Spore
Toxin
Potentially hazardous food
Temperature Danger Zone

Time-Temperature history
Bacterial growth curve
Cross-contamination
Anaerobic
Aerobic
Virus
Fungi
Cumulative effect
Staphylococcus aureus

Clostridium botulinum
Clostridium perfringens
Salmonella sp.
Shigella
Bacillus cereus
Wholesomeness
FDA Model Ordinance
Sanitization

EXERCISE 1: FACTORS AFFECTING THE MICROBIAL SAFETY OF FOODS

A. Sources of Contamination

PROCEDURE

1. Based on readings, list common sources of food product contamination (e.g., infected food handlers, chemicals).
2. Identify methods by which contamination may be introduced into foods.

Sources of Contamination	Methods of Food Contamination
1.	
2.	
3.	
4.	
5.	

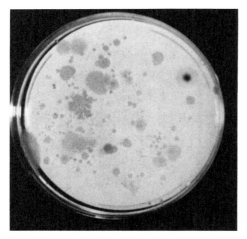
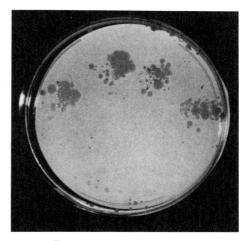

BACTERIAL GROWTH FROM CONTAMINATED APRON. **BACTERIAL GROWTH FROM UNWASHED HANDS.**

Courtesy: University of Georgia, College of Agriculture

B. CONDITIONS NECESSARY FOR THE GROWTH OF BACTERIA

PROCEDURE

1. Identify environmental conditions that bacteria need in order to grow.

Environmental Conditions Necessary for Bacterial Growth

2. List examples of foods that support microbial growth (called "potentially hazardous foods" in state and local food codes).

40 DIMENSIONS OF FOOD

C. BACTERIAL GROWTH CURVE

Contaminated foods kept at unsafe temperatures, e.g., 45°F to 140°F (7°C to 60°C) in the temperature danger zone become unsafe to eat after a period of time (2 to 4 hours).

PROCEDURE

1. Study the theoretical growth curve of bacteria.
2. Using references, identify what is occurring during each phase of bacterial growth.

Phase Name and Description

(a) A–B: LAG

(b) C–D: LOG

(c) E–F: STATIONARY

(d) G–H: DECLINE

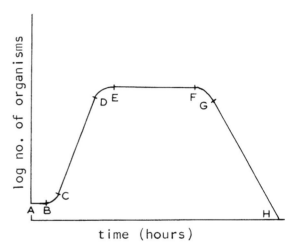

GROWTH CURVE OF MICROORGANISMS
Source: Frazier WC, 1967. Food Microbiology. McGraw-Hill Book Co. Reprinted with permission.

1. Why is there a lag in growth before numbers of bacteria begin to increase?

2. What would be the effect of refrigerating a food during the lag phase? Does *refrigeration* eliminate the effect of the time already spent in the lag phase? Does *freezing* kill microorganisms?

3. How would microbial growth be affected if a refrigerated, "potentially hazardous food" were again held at room temperature, this time for several hours?

SANITARY DIMENSIONS 41

4. Explain why it is important to know the time-temperature history of foods that are to be served.

5. By example, explain what is meant by the cumulative effect as it operates with a potentially hazardous food.

6. Summarize the conditions that must exist if an illness is to result from the ingestion of a food.

7. Explain practical control measures that should be employed to prevent foodborne illnesses.

EXERCISE 2: TEMPERATURE CONTROL IN FOOD HANDLING

A. FACTORS AFFECTING THE RATE OF COOLING OF LARGE QUANTITIES OF FOODS

PROCEDURE

1. Prepare 4 qt (or 4 L) of a boiling, thickened liquid mixture.[1]
2. Divide the liquid into four containers, using two oblong pans and two tall 1-quart (1-L) glass measures. Record the initial temperature of the product.
3. Place one of each size container in a 40°F (4.4°C) refrigeration unit, and let the other two containers cool at room temperature.
4. Record temperatures at designated times.

	Temperature			
	Initial	10 min	20 min	30 min
Cooled at room temperature a. 1 qt (L) measure b. oblong, shallow pan				
Cooled under refrigeration a. 1 qt (L) measure b. oblong, shallow pan				

[1] For laboratory purposes, use quart or liter glass measuring cup.

42 DIMENSIONS OF FOOD

1. How does the speed of cooling a hot food adversely affect the sanitary quality of that food?

2. How does the density or viscosity of the food material affect the rate of cooling?

3. In order to ensure rapid cooling and proper storage of perishable foods, what temperature range is recommended for refrigeration?

B. TEMPERATURES FOR HOLDING AND REHEATING FOODS

PROCEDURE
1. Complete the chart, listing temperatures required for cold and hot holding of "potentially hazardous foods" (this may vary by state, county, or city jurisdiction).
2. Record temperature requirements for reheating of potentially hazardous foods.

Cold holding	
Hot holding	
Reheating	

C. RECOMMENDED TEMPERATURES FOR COOKED FOOD

PROCEDURE
1. Complete the chart regarding recommended final cooking temperatures to ensure food wholesomeness.
2. List rationale for the temperature requirements of products.

Food Product	Temperature	Rationale
Eggs		
Poultry		
Pork		
Roast beef		
Steak		
Ground beef		

EXERCISE 3: SANITIZATION IN THE FOOD PREPARATION ENVIRONMENT

A. Use of Approved Chemical Sanitizers

PROCEDURE

1. List three chemical sanitizing agents approved by the Environmental Protection Agency for use on food contact surfaces.
2. List concentrations of sanitizers (parts per million) required for sanitizing equipment and utensils by immersion (in a sink), and for sanitizing equipment in-place (equipment that is too large to fit in the sink, or that is electrically based and must be cleaned and sanitized in place).

Chemical Name	Concentration: Immersion	Concentration: In-place Equipment
1.		
2.		
3.		

B. Sanitization by Immersion

PROCEDURE

1. Identify the correct arrangement of a three-compartment sink and the function of each sink compartment.

Commercial/Institutional Sink	Function
First compartment sink	
Second compartment sink	
Third compartment sink	

2. Specify, in the table below, the two sanitizing methods that are effective in the sanitizing sink.

Sanitizing Method	Length of Exposure Time	Water Temperature
1.		
2.		

SUMMARY QUESTIONS—SANITARY DIMENSIONS

1. Differentiate between *spoiled* and contaminated or *unwholesome* food products. Provide examples.

2. Why is it possible for a food to be unwholesome but not spoiled? May signs of spoilage also be indications of unwholesomeness?

3. Describe legal protection that consumers have against unwholesome food.

4. Define foodborne illness. Discuss additional, emerging pathogens, not listed in Appendix G, that need to be controlled in order to prevent foodborne illness.

5. Define and distinguish between intoxication and infection.

6. Explain cross-contamination by an example.

7. While commercially prepared mayonnaise (pH 3.0 to 4.1) is not a potentially hazardous food, mayonnaise-based salads (e.g., potato salad or tuna salad) are often causes of foodborne illness. Explain. Specify what ingredients, cooking procedures and environmental factors make such foods potentially unwholesome.

8. If you, or others, associate your illness with food, to whom should the incident be reported?

9. According to the Centers for Disease Control and Prevention (CDC), roast beef and poultry are frequently reported in foodborne illness cases.
    a. Explain why this is true: consider cooking and holding temperatures and the composition of the food.

    b. Which microorganisms are most frequently the cause of illness?

10. The following information is adapted from a case history report of the Communicable Disease Center.

> On March 10, 64 cases of acute gastrointestinal illness occurred among 107 guests shortly after eating at a wedding reception. The reception food was prepared in private homes and then brought to the reception. Specifically, 40 chickens were cooked and deboned on the 8th, then refrigerated overnight. On the 9th, the meat was ground in a meat grinder with celery and onions. Then, the mixture was mixed with mayonnaise and refrigerated. On the day of the reception (the 10th), the salad was not refrigerated en route to the reception or during the reception.

Comment on this case, noting specific problems connected with the procedures that could be expected to cause illness of guests.

11. List the four major factors or situations that must be studied to determine the causes of any foodborne illness. Be specific in your answer.

12. A fictitious news article reads as follows:

> **75 Children Ill**
> **From Picnic Food**
> SANDY POINT, U.S.A.—Poisonous food transformed a gala school picnic into mass misery Saturday night. About 75 children from Sandy Point Central School fell violently ill at their annual end-of-school picnic. Children fell ill about 3 hours after eating a delicious picnic supper.

   a. What kind of illness would you suspect?

   b. What foods might be involved?

   c. Why is a picnic food often the cause of illness?

13. Identify the "Safe Handling Instructions" that appear on packaged meat and poultry. What is the reason for conveying this information to consumers?

F. Food Processing Dimensions

OBJECTIVES

To recognize temperatures commonly used in food processing
To identify how processing temperatures are influenced by type of processing equipment
To distinguish characteristics of standard equipment and processes used in canning and freezing
To use, compare and evaluate common methods of canning acid and nonacid foods
To use, compare and evaluate common methods of freezing fruits and vegetables

REFERENCES

Appendices F, G, H
USDA
Cooperative Extension

ASSIGNED READINGS

TERMS

Simmer	Hermetic seal	Cold pack	Toxin
Scald	pH	Head space	Spore
Boil	Low acid	Blanch	*Clostridium botulinum*
Pressure cook	Acid	Syrup pack	Flat sour
Pressure canner	Open kettle	Dry sugar pack	Shelf life
Pressure sauce pan	Water bath	Enzymatic browning	Freezer storage life
Petcock	Hot pack	Venting, exhausting	USDA

EXERCISE 1: PROCESSING TEMPERATURES

PROCEDURE

1. Place 2 c (480 ml) of water in each container listed below.
2. Cover containers with lids and regulate heat to maintain temperature.
3. After 10 minutes, check and record temperature of water bath.

Container	Temperature after 10 min
1. Covered saucepan (water boiling)	
2. Steamer	
3. Water in top of double boiler (over simmering water)	
4. Water in top of double boiler (over boiling water)	
5. Water in top of double boiler (surrounded by boiling water)	
6. Pressure saucepan (demo) (15 pounds pressure)	

PRESSURE CANNER
Courtesy: USDA

1. Account for the differences in final temperature:
   a. Covered saucepan or steamer and pressure cooker at 15 pounds pressure

   b. Top of double boiler, surrounded by boiling water; top of double boiler, over boiling water

2. At what temperature does water "simmer"? Describe the appearance of the water.

3. Explain what happens when water boils. Describe the appearance of the water.

4. Compare temperature of boiling water and temperature of steam.

5. When will water boil at temperatures above 212°F (100°C)? Why?

6. When will water boil below 212°F (100°C)? Why?

7. How does pan shape and use of lid influence rate of water evaporation during cooking?

8. Discuss the rate of heat transfer by conduction, convection and radiation. Give an example of each method as used in food preparation.

EXERCISE 2: FOOD PROCESSING, CANNING

A. Canning Equipment

PROCEDURE
1. Examine the different kinds of jars and closures available for canning. Note how closures attach to jars.
2. Examine the following processing equipment and note characteristics:
    a. Pressure canner
    b. Boiling water bath
    c. Open kettle

ONE TYPE OF CANNING JAR
Courtesy: USDA

B. CANNING ACID AND LOW-ACID FOODS

PROCEDURE

1. Prepare and process 1 pint (480 ml) of vegetable (low acid) or 1 pint (480 ml) of fruit (acid). Follow current directions provided by the USDA.
2. When cool enough to handle, label jar with processing method used, name, and date.
3. In a subsequent laboratory, examine and evaluate canned products as to color, texture and flavor. Complete the table below and summarize your conclusions.

Product	Relative Acidity	Canning Method	Processing		Palatability		
			Time	Temp	Color	Texture	Flavor

Conclusions:

QUESTIONS—CANNING

1. Why is the allowance of a "head space" important in packing jars for canning?

2. What foods can be canned in an open saucepan? What are the restrictions of using this method? Why must low-acid foods be canned in a pressure canner?

3. Must jars be sterilized when the boiling water bath or pressure cooker are used? Explain.

4. In using the pressure canner:
    a. How and why is the canner exhausted?

    b. When is the processing time counted?

    c. Why must the pressure return to zero prior to opening the petcock and removing the cover?

5. In using the water bath method:
    a. What temperature should the water bath be when jars are put into the bath? Why?

    b. What is the height of water in the pan, relative to the jars? Why is this important?

    c. How is "processing time" counted?

6. How are processed jars tested for a complete seal?

7. Why are screw bands removed after jars are cold and sealed?

8. What factor(s) cause "flat sour"? Suggest ways this can be avoided.

9. Why are up-to-date references on canning essential?

EXERCISE 3: FOOD PROCESSING, FREEZING

A. Freezing Equipment

PROCEDURE

Examine the various kinds of rigid and nonrigid freezing containers available, e.g., glass jars, plastic boxes, waxed cardboard, bags, and sheets of moisture-resistant cellophane, foil, pliofilm, polyethelene

B. Freezing Fruits and Vegetables

PROCEDURE

1. Freeze 1 pint (480 ml) of assigned fruit or vegetable, following current directions provided by the USDA, except for pretreatment indicated on chart below.
2. Label containers and freeze.
3. In a subsequent laboratory, examine and evaluate all products. Summarize your conclusions.

Product	Pretreatment	Type of Container	Evaluation of Thawed Product
Fruit	Packed in syrup, no ascorbic acid		
Fruit	Packed in syrup, ascorbic acid		
Fruit	Dry sugar pack, ascorbic acid		
Vegetable	Blanched		
Vegetable	Unblanched		

Conclusions:

QUESTIONS—FREEEZING

1. What is the function of ascorbic acid when it is added to syrup for fruits? Which fruits need ascorbic acid to ensure a palatable product?

2. Explain the role of blanching in freezing foods.

3. Why are vegetables chilled immediately after blanching?

4. Why must head space be allowed in freezer containers?

BLANCHING VEGETABLES PRIOR TO FREEZING
Courtesy: USDA

5. What freezer temperature is recommended for freezing and storing frozen foods?

6. How long will frozen fruits and vegetables maintain high palatability characteristics in the freezer?

7. Does freezing kill *Clostridium botulinum* spores? Explain.

8. What foods generally freeze well? What foods do not?

9. What guidelines can be used to evaluate the wholesomeness of a frozen food?

SUMMARY QUESTIONS—DIMENSIONS OF FOOD

1. Many say variety in food choices is a basic key to good nutrition. Do you agree? Why or why not?

2. You are an Action worker in South America working with adolescent girls. You hear that an American company has developed Fe-ol, a new, carbonated, iron-enriched beverage. Based on criteria for food selection, list several specific factors that you would consider and investigate before recommending the Fe-ol be used in your program.

3. With rising prices, many consumers are changing food buying habits. Discuss some ways that money can be saved on food purchased without compromising nutritive value or personal preferences.

4. As a nutritionist working in an underdeveloped country, you have surveyed the diets of the population and concluded that the people there were not meeting their RDAs for vitamin A, calcium, and iron. This lack was especially true for women under age 50 and children. Your challenge is to devise and implement a basic food guide that will help improve the nutritional status of the population. After studying the food supply and dietary customs, you ascertain the following facts:

   - Most people are already eating ample quantities of rice, coffee, starchy vegetables (plantains, green bananas), lard, sugar, dried imported codfish.
   - Imported canned fruits are considered prestige foods.
   - Native citrus fruits are plentiful but are not considered prestige foods.
   - Deep-yellow fruits (papaya, mango) and vegetables (squash, sweet potato) are plentiful.
   - Dried beans can be imported at a reasonable price.
   - Eggs are fairly plentiful, often produced at home.
   - Although an island nation, fish, other than dried cod, is not plentiful.
   - Meat is high priced, since it must be imported.
   - Because of a lack of refrigeration, some areas of the island do not have access to fresh milk.

   Briefly discuss how you would devise an appropriate food guide. Give reasons for your choices of food groups.

54 DIMENSIONS OF FOOD

5. Indicate on the thermometer the temperatures for the following situations:
   a. temperature at which water freezes
   b. optimum refrigerator temperature
   c. optimum freezer temperature
   d. temperature of simmering water
   e. temperature of food cooking in top of double boiler held over *simmering* water
   f. temperature of food cooking in top of double boiler held over *boiling* water
   g. cooking at 15 lb pressure
   h. *Salmonella* destroyed
   i. *Staphylococcus* destroyed
   j. *C. perfringens* destroyed
   k. *C. botulinum* destroyed
   l. toxin of *C. botulinum* destroyed
   m. toxin of *Staphylococcus* destroyed
   n. temperature for canning *low*-acid foods
   o. temperature for canning *acid* foods

6. Bracket temperature zone of most rapid growth of microorganisms.

7. Bracket temperature zones that prevent rapid growth of microorganisms but allow their survival.

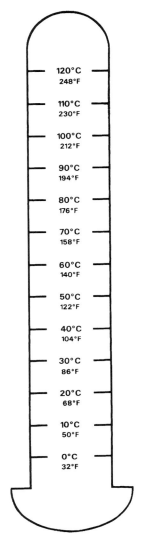

8. Keep an accurate log of all the foods you consume on a weekend. From this log, identify snacks and evaluate their nutritive value.
   a. How many calories did your snacks add to your daily intake?

   b. Did any of the snacks add important nutrients you needed to meet your RDA?

9. Compile a checklist to use in evaluating the sanitation practices of a restaurant or lunch counter. Visit an eating place, especially one in which you, the customer, are able to observe kitchen activities and test the validity of your checklist. Based on your observations, if you were the manager, what specific suggestions would you make to your staff concerning improved handling of foods?

10. Based on your study of the economic dimensions of food, what suggestions would you make to food companies or to the government, for improvements that would enable shoppers to make better buys?

11. Why are directions for processing foods sometimes revised?

12. What general advice would you give to anyone who plans to home-process food?

PART II

Food Principles

Acceptance and enjoyment of food are largely a matter of "eating quality." Application of the principles of food preparation enhances eating quality and maximizes nutritional value. This section is designed to help you understand and apply these principles.

Functional and structural properties of food constituents and their behavior in food preparation are emphasized. Basic principles are demonstrated in a series of experiments and, then, applied in the preparation of various food products. Experiences with all major food groups are included. Nutritional value of these foods is emphasized.

Through questions and problems, your understanding of all the dimensions of food can be applied to specific food groups. From the overall experiences in this section, you will learn and be able to predict, how preparation affects and changes food, not only in terms of palatability, but also in nutritive value and sanitary quality.

Source: Division of Nutritional Sciences, New York State College of Human Ecology at Cornell.

A. Measurements, Use of Ingredients, and Laboratory Techniques

OBJECTIVES

To become familiar with common techniques of food preparation and cleanup
To delineate utensils and methods for measuring liquids and solids
To measure liquids and solids accurately
To become familiar with metric equivalents

REFERENCES

Appendices I, K

ASSIGNED READINGS

TERMS

Sift   Pare
Beat   Fliud ounces
Chop   Metric equivalent
Fold   Meniscus

MEASURING ACCURATELY
Source: Division of Nutritional Sciences, New York State College of Human Ecology at Cornell.

USE OF SALT AND FAT

Seasoning: Salt is not included in recipes, except for yeast bread. Season to taste and limit added salt. Use of herbs and spices is encouraged. See Appendix N.

Use of fat: For margarine, select highly polyunsaturated brands; for oils, polyunsaturated or monounsaturated brands are suggested. Use of no-stick cooking sprays is suggested in place of greasing pans or casseroles.

USE OF OVENS

In the interest of energy conservation, ovens should not be set until necessary. Instructions to set the oven are provided at the beginning of recipes to serve as a reminder to set the oven.

Refer to Microwave Cooking chapter for microwave oven recipes and to Appendix H for general information on heat transfer.

EXERCISE 1: DEMONSTRATION OF MEASURING AND MIXING TECHNIQUES

Procedure

1. Observe a demonstration of measuring techniques, noting equipment for dry and liquid measures.
2. Calculate values and complete the following chart:

1 cup liquid measure	=	fluid ounces
	=	milliliters (ml)
	=	liters
1 cup dry measure	=	Tablespoons (T)
1 T	=	teaspoons (t)
	=	ml
1 t	=	ml
1 lb all-purpose flour	=	cups (c)
	=	g
1 lb granulated sugar	=	c
	=	g
1 lb butter/margarine	=	c
	=	T
	=	g
4 oz cheddar cheese	=	g
3 oz hamburger	=	g

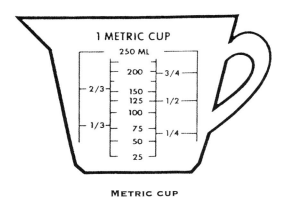

METRIC CUP

Source: Division of Nutritional Sciences, New York State College of Human Ecology at Cornell.

EXERCISE 2: MEASURING LIQUIDS

Procedure

1. Select the appropriate measuring utensils.
2. Place container on level surface. Carefully pour liquid into container. Eye should be level with the measure mark and liquid added until bottom of meniscus rests on mark.

EXERCISE 3: MEASURING SOLIDS

Procedure

1. Weigh the appropriate measuring utensil on an ounce scale.
2. For dry solids (e.g., granulated sugar, flour), lightly spoon material into container. For moist solids (e.g., brown sugar, solid fats), firmly pack the material into container.
3. Reweigh filled container on scale; record actual weight of material.
4. Convert to metric equivalents; record on table.
5. Compare and discuss variability of weights recorded by class members for same measure.

Measurements, Ingredients, and Laboratory Techniques 59

	Measure	Weight oz	Weight g
Sugar	1 t (5 ml)		
	1 T (15 ml)		
	1 c (240 ml)		
Margarine	1 t (5 ml)		
	1 T (15 ml)		
	1 c (240 ml)		

EXERCISE 4: CLEAN-UP

PROCEDURE

Observe procedures to be followed in clean-up of utensils, stoves and kitchen unit. Consider appropriate disposal of waste to facilitate recycling.

SUMMARY QUESTIONS—MEASUREMENTS, USE OF INGREDIENTS, AND LABORATORY TECHNIQUES

1. Given ⅔ c (160 ml) hydrogenated shortening, what type of measuring equipment should be used?

2. How does the technique for measuring solid fat differ from such solid ingredients as baking powder and sugar?

3. When measuring 5 T (75 ml) of milk, what equipment would be most accurate?

4. In measuring ⅜ c (90 ml) flour, what equipment should be used?

5. In a recipe that calls for 3 c (720 ml) sugar, how much (in pounds, grams) will need to be purchased?

6. Regarding stick margarine:
    a. What is the easiest way to measure ¾ c (180 ml)? ¼ c (60 ml)?

    b. ½ c (120 ml) margarine is equal to how much of a pound? How many grams?

7. A recipe requires 7 c (1680 ml) all-purpose flour, ¾ lb (340.5 g) is available. How much more will need to be purchased? in pounds? in grams?

B. Cereal and Starch

OBJECTIVES

To define and explain the role of separating agents in starch cookery

To describe the events that occur in starch cookery, their relationship to temperature, thickness and flavor of the product

To delineate the effect of sugar and acid on starch thickened products

To recognize individual properties of flour and cornstarch

To relate principles of starch cookery to the preparation of cereal products

To relate principles of starch cookery to a variety of starch thickened products

To prepare a palatable starch product, delineating and giving reasons for each step

To appraise the nutritive, sanitary, and economic dimensions of starch and cereal products

REFERENCES

ASSIGNED READINGS

TERMS

Bran	Polymer	Separating agent	Colloidal dispersion
Germ	Amylose	Adsorption	Sol
Endosperm	Amylopectin	Absorption	Gel
Enriched	Granule	Imbibition	Gelation
Starch	Suspension	Gelatinization	Viscosity
Modified starch	Maltodextrins		

STANDARD PROCEDURE
for Making Flour/Cornstarch Thickened Product
(reference for Exercises 1–4)

1. In a saucepan, mix the starch and separating agent (melted fat, sugar, cold liquid).
2. *Slowly* add remaining liquid, stirring constantly.
3. Cook, stirring constantly until mixture boils.
4. Cook a few minutes longer to improve flavor, stirring gently.

PALATABILITY TERMS			
Texture	Consistency	Appearance	Flavor
Lumpy	Thin	Opaque	Raw
Fairly smooth	Medium thick	Cloudy	Cooked
Smooth	Thick	Translucent	
	Gel-like	Transparent	

EXERCISE 1: SEPARATION OF STARCH GRANULES

PROCEDURE

1. Prepare a starch thickened product following the STANDARD PROCEDURE and using the proportions of ingredients listed below.
2. Evaluate consistency of hot products.

Starch	Separating Agent	Boiling Water	Texture-Consistency of Hot Sauce
1 T (15 ml) flour	None	½ c (120 ml)	
1 T (15 ml) flour	1 T (15 ml) melted fat	½ c (120 ml)	
1 T (15 ml) flour	¼ c (60 ml) cold water	¼ c (60 ml)	
1 T (15 ml) flour	1 T (15 ml) sugar	½ c (120 ml)	

1. What is the scientific explanation for lump formation?

2. Explain how the following function as separating agents:

   a. Melted fat

   b. Cold liquid

   c. Sugar

EXERCISE 2: PROPERTIES OF WHEAT AND CORNSTARCH

PROCEDURE

1. Prepare a starch thickened sauce with each of the starches listed by following the STANDARD PROCEDURE. Use ¼ c (60 ml) cold water as the separating agent, then add ¾ c (180 ml) water.
2. Record observations as to consistency and appearance of sauces while hot and after cooling.
3. Save the hot and cool sample of the cornstarch mixture to use as the *Standard* in Exercise 3.

Starch	Observations	
	Hot Sauce	Cooled Sauce
2 T (30 ml) cornstarch (STANDARD)		
2 T (30 ml) flour		
2 T (30 ml) browned flour[1]		

[1]To brown, spread flour in a thin layer in a flat pan. Bake at 375°F (190°C), stirring frequently, until flour is light brown.

1. Draw a diagram of:

    a. Starch in cold water

    b. Starch–water mixture heated to boiling

    c. Cooked mixture cooled to refrigeration temperature

2. Explain, using scientific terms, how a starch gel is formed from a starch sol.

3. What factor(s) determine the type of separating agent to be used in a starch-thickened product?

4. How could the same thickness in a final product be obtained if a recipe listed 1 T (15 ml) cornstarch and only flour was available?

5. Why is a flour product cooked for additional time after maximum thickness is reached?

EXERCISE 3: EFFECT OF SUGAR AND ACID ON GELATINIZATION

PROCEDURE

1. Prepare a starch thickened sauce following the STANDARD PROCEDURE, with 2 T (30 ml) cornstarch, and using the proportions of ingredients listed below. (Cook the vinegar mixture slowly.)
2. Evaluate consistency of hot and cooled sauces compared to the STANDARD in Exercise 2.
3. Record observations as to consistency of hot and cold products.

Separating Agent	Water	Observations	
		Hot Sauce	Cooled Sauce
½ c (120 ml) sugar	1 c (240 ml)		
¼ c (60 ml) vinegar	¾ c (180 ml)		
¼ c (60 ml) water	½ c (120 ml) + ¼ c (60 ml) vinegar (added after thickening)		

1. Explain how large amounts of sugar affect the gelatinization of starch and characteristics of the sol/gel.

2. Explain how acid affects the gelatinization of starch. How does the time of acid addition affect product results?

3. Based on these experiments, when should lemon juice be added to a lemon pie filling mixture?

EXERCISE 4: APPLICATION OF PRINCIPLES TO STARCH-THICKENED PRODUCTS

PROCEDURE

1. a. Prepare a starch thickened gravy (1 c, 240 ml) from the following ingredients: beef bouillon, flour and hydrogenated fat.
   b. Show gravy to instructor.

   AND/OR

2. a. Prepare a starch thickened product from the assigned list of ingredients. Based on previous experiments and readings, outline procedures, giving scientific reasons for each step.
   b. Display product.

Product: Separating Agent Used:

Steps	Explanation

3. Evaluation: Analyze success or failure of the product, based on scientific principles.

	Explanation
Texture:	
Consistency:	

4. Evaluate all products, noting types of separating agents used and general palatability.

Recipe Ingredients[1]	Evaluation
CHEESE SAUCE 1 c 240 ml milk ⅓ c 80 ml processed cheese dash dash paprika 2 T 30 ml flour 2 T 30 ml margarine	
CORN CHOWDER 2 T 30 ml finely chopped onion 1 c 240 ml milk 2 T 30 ml margarine 1 1 chicken bouillon cube 1 T 15 ml flour ½ c 120 ml creamed corn	
CREAM OF POTATO SOUP 1½ c 360 ml milk ½ T 7.5 ml pimento, chopped 1 T 15 ml margarine 1 c 240 ml diced cooked potatoes 1 T 15 ml flour	
RAISIN SAUCE FOR MEAT ½ c 120 ml brown sugar 2 T 30 ml lemon juice 1 t 5 ml dry mustard[2] ¼ t 1.25 ml grated lemon rind 2 T 30 ml flour 1½ c 360 ml water 2 T 30 ml vinegar ⅓ c 80 ml raisins	
TOMATO SAUCE 1 T 15 ml green pepper, chopped 1 T 15 ml cornstarch 1 t 5 ml grated onion 1 c 240 ml tomato juice 1 T 15 ml margarine	

[1]Circle separating agent used.
[2]Treat mustard as starch.

Cereal and Starch 67

Recipe Ingredients[1]							Evaluation
<td colspan="7" align="center">FRUIT SAUCE</td>							
2 T	30 ml	margarine	2 T	30 ml	lemon juice		
¾ c	180 ml	confectioner's sugar	2 t	10 ml	orange rind		
2 T	30 ml	cornstarch	⅓ c	80 ml	orange juice		
<td colspan="7" align="center">LEMON SAUCE</td>							
1 T	15 ml	cornstarch	¼ c	60 ml	lemon juice		
¾ c	180 ml	water	¼ t	1.25 ml	grated lemon rind		
⅓ c	80 ml	sugar	dash	dash	nutmeg		
1 T	15 ml	margarine					
<td colspan="7" align="center">PUDDING, BUTTERSCOTCH[2]</td>							
1 c	240 ml	milk	1 T	15 ml	margarine		
⅓ c	80 ml	brown sugar, packed	2 T	30 ml	cornstarch		Microwave:
½ t	2.5 ml	vanilla					
<td colspan="7" align="center">PUDDING, CHOCOLATE[2]</td>							
¼ c	60 ml	sugar	½ t	2.5 ml	vanilla		
1½ T	22.5 ml	cornstarch	½ sq	½ sq	unsweetened		Microwave:
1 c	240 ml	milk			chocolate, grated		
1 T	15 ml	margarine					
Chocolate Pudding (instant)							
Pregelatinized starch							
Chocolate Pudding (Canned)							
High amylopectin starch							
<td colspan="7" align="center">SWEET-SOUR SAUCE</td>							
½ c	120 ml	pineapple juice	½ t	2.5 ml	prepared mustard[3]		
1 T	15 ml	vinegar	1 T	15 ml	cornstarch		
2 T	30 ml	brown sugar	¼ c	60 ml	water		
½ t	2.5 ml	paprika	½ c	120 ml	pineapple tidbits		
1	1	chicken bouillon cube	¼ c	60 ml	green pepper, chopped		

[1]Circle separating agent used.
[2]See Microwave Cooking chapter for microwave recipes.
[3]Treat mustard as starch.

CEREALS

RICE

Source: Cornell University, Cooperative Extension.

EXERCISE 5: PREPARING CEREAL PRODUCTS

PROCEDURE

1. Cook ¼ c (60 ml) cereal product as directed, unless otherwise assigned.
2. Measure, record yield, and display.
3. Record palatability assessments of all products.

PALATABILITY TERMS		
Texture	Consistency	Flavor
Smooth	Thick	Sweet
Lumpy	Thin	Nutty
Grainy	Gel-like	Raw Starch
	Gummy	
	Sticky	

COOKING DIRECTIONS FOR CEREALS/GRAIN PRODUCTS:

Granular or finely milled grains—cornmeal, farina, grits. Add grain to cold water. Boil gently. Stir occasionally.

Whole or coarsely milled grains—barley, buckwheat (Kasha), bulgur, rice, oats. Add grain to boiling water. Boil gently. Add water as needed.

Pasta—Add to boiling water. Boil uncovered, until desired tenderness is achieved. To prevent boiling over, add ½ t (2.5 ml) oil to cooking water.

Evaluation of Cereal Products

Grain ¼ c (60 ml)	Water (c, ml)	Cooking Time (min)	Cooked Yield	Palatability Characteristics
GRANULAR Farina	1½ c (360 ml)	3–5		
Cornmeal	1½ c (360 ml)	3–5		
Cornmeal grits	¾ c (180 ml)	5		
FLAKED Oatmeal, regular	¾ c (180 ml)	15		
Oatmeal, quick	¾ c (180 ml)	2		
Oatmeal, quick	¾ c (180 ml)	2 min, stir while cooking; 5 min more		
Oatmeal, instant package	pkg directions			
WHOLE Rice, regular	½ c (120 ml)	15–20		
Rice, converted	½ c (120 ml)	15–20		
Rice, brown	½ c (120 ml)	40		
Rice, instant	¼ c (60 ml)	boiling water over rice, stand 5 min		
OTHER GRAINS[2] Wheat Bulgur	½ c (120 ml)	15		
Wheat Bulgur	½ c (120 ml)	cold water, soak 10 min		
Barley, qk cooking	½ c (120 ml)	15		
Buckwheat (Kasha)	½ c (120 ml)	15		
Buckwheat (Kasha) ½ c	pkg directions use ½ egg			
Couscous (semolina)	¼ c (60 ml) boiling water	soak 5 min		
PASTA Pasta	1 pt (480 ml)	7–12		
Whole wheat	1 pt (480 ml)	7–12		

[1]With long-cooking grains, e.g., brown rice, additional water may have to be added during cooking.
[2]Experiment with other grains, e.g., millet and quinoa (Keenwa).

CEREAL RECIPES

Barley Cheese Casserole

water	3 c	720 ml	margarine	2 T	30 ml
quick pearled barley	1 c	240 ml	cooked tomatoes, drained	2 c	480 ml
finely chopped onion	¼ c	60 ml	cheese, grated	2 oz	57 g

1. Set oven at 350°F (175°C).
2. Add barley to boiling water. Cover and simmer 10 to 12 minutes until barley is tender. Stir occasionally. Drain.
3. Sauté onion in fat until tender.
4. Add onion and remaining ingredients to greased casserole.
5. Bake 10 to 15 minutes until casserole is heated through or reheat in a frying pan on top of stove. Season. (3 to 4 servings)

Cheesy Corn Grits

quick-cooking grits	½ c	120 ml	milk	⅓ c	80 ml
milk	1½ c	360 ml	cheese, grated	½ c	120 ml
egg, slightly beaten	1	1	chives	2 T	30 ml

1. Boil grits in milk.
2. Add remaining ingredients and place in small greased casserole. Bake at 350°F (175°C) 35 to 40 minutes. (2 to 3 servings)

Fiesta Rice

water	2 c	480 ml	chopped pimento	1 T	15 ml
rice	1 c	240 ml	chili powder	¼ t	1.25 ml
margarine	1 T	15 ml	drained cooked tomatoes	½ c	120 ml
chopped green pepper	2 T	30 ml			

1. Cook rice in unsalted water about 18 minutes. Drain and keep hot.
2. Sauté pepper and pimento in fat until softened. Stir in chili powder and tomatoes.
3. Combine hot vegetable mixture and rice. Reheat. (2 servings)

Macaroni and Cheese

macaroni	½ c	120 ml	dry mustard	⅛ t	.63 ml
chopped onion	¼ c	60 ml	pepper	⅛ t	.63 ml
margarine	1 T	15 ml	milk	1 c	240 ml
flour	1 t	5 ml	processed American Cheese	1 c	240 ml

1. Cook macaroni in large amount of water until just tender. Drain.
2. In a medium saucepan, sauté onion until tender. Add flour and seasonings. Stir in milk.
3. Cook over medium heat, stirring until thickened. Reduce heat; add cheese and macaroni. Stir. Let stand 5 minutes (2 servings).

Variation: Add 1 c (240 ml) cooked vegetable with cheese.

Cereal and Starch

Pasta Primavera

thin noodles	1 c	240 ml	garlic powder	dash	dash
celery, thinly sliced	¼ c	60 ml	vegetable oil	½ t	2.5 ml
green beans, diced	¼ c	60 ml	green peas, frozen	½ c	60 ml
carrots, diced	¼ c	60 ml	flour	1 t	5 ml
red onion, sliced	¼ c	60 ml	margarine	½ t	2.5 ml
green pepper strips	1 T	15 ml	milk	⅓ c	80 ml
basil leaves	¼ t	1.25 ml			

1. Cook noodles according to package directions. Drain.
2. Stirfry fresh vegetables and seasonings in oil, lifting and turning.
3. Add frozen peas, cover and reduce heat. Cook for about 2 minutes until vegetables are tender crisp. Remove vegetables from pan.
4. Mix flour and margarine in fry pan. Add milk slowly, heat, stirring constantly, until thickened. (Sauce will be thin.)
5. Add sauce and vegetables to noodles, mix gently. Heat to serving temperature. (2 servings)

Pasta Salad

pasta, spirals	4 oz	114 g	onion, minced	2 T	30 ml
broccoli flowerets	½ c	120 ml	garlic, minced	1 t	5 ml
carrots, diced	¼ c	60 ml	Italian dressing	¼ c	60 ml
black olives, sliced	½ c	120 ml	Dijon mustard	½ t	2.5 ml

1. Boil pasta as directed on package. Drain and allow to cool.
2. Steam broccoli and carrots until tender crisp. Cool.
3. Combine all ingredients, tossing well. Refrigerate. (2 to 4 servings)

Polenta

coarse cornmeal	½ c	120 ml	paprika	⅛ t	.63 ml
water, boiling	2 c	480 ml	grated cheese	¼ c	60 ml

1. Slowly add cornmeal to boiling water.
2. Cook, stirring frequently, over low heat about 15 minutes, or until thickened (this may be cooked in the top of a double boiler or over boiling water).
3. Pour hot mixture into 7-inch (18-cm) pie pan or small cake pan.
4. Sprinkle with paprika and cheese. Cover and refrigerate. If desired, reheat before serving. Note: Serve plain, top with spaghetti or pizza sauce or sautéed green peppers and onions. (3 servings)

Rice Pudding

cooked rice	1 c	240 ml	vanilla	½ t	2.5 ml
milk	¾ c	180 ml	lemon peel	dash	dash
sugar	2 T	30 ml	lemon juice	½ t	2.5 ml
egg, slightly beaten	1	1	raisins	2 T	30 ml

1. Set oven to 325°F (165°C).
2. Combine all ingredients and pour into greased casserole.
3. Bake 30 to 35 minutes. (3 to 4 servings)

Tabouleh

bulgur	½ c	120 ml	finely chopped, seeded cucumber	½ c	120 ml
cold water	1 c	240 ml	finely chopped bell pepper	½ c	120 ml
finely chopped tomato	1	1	fresh lemon juice	⅓ c	80 ml
finely chopped parsley	1 c	240 ml	olive oil	2 T	30 ml
finely chopped scallions	½ c	120 ml			

1. Place bulgur and water in a bowl, and refrigerate 1 hour. Drain if necessary.
2. Combine remaining ingredients. Add to bulgur, stirring to blend. Refrigerate.
3. Before serving, stir salad and adjust seasoning. (6 servings)

EVALUATION—CEREAL RECIPES

Cereal Recipe	Palatability

CEREAL AND STARCH

BULGUR

COUSCOUS

Courtesy: Wheat Foods Council

SUMMARY QUESTIONS—CEREAL AND STARCH

STARCHES

1. A fruit sauce recipe calls for 3 T (45 ml) cornstarch. How much sugar would be needed as a separating agent? How much fat? How much cold liquid?

2. When preparing a sauce or gravy, when will maximum thickness be observed? How can thickness of hot gravy be increased if it is necessary?

3. How does the starch in "instant" pudding differ from the starch used in "homemade" forms?

4. How are modified starches and maltodextrins used in the food industry? Explain.

5. Why are creamed (*milk* thickened sauce) foods and *meat gravies* frequently considered a health risk?

CEREALS

1. What are the desired palatability characteristics of a cooked grain product?

2. What is the major problem encountered in cooking finely milled grains to achieve a smooth texture? How is this problem solved?

3. Why do cereal products expand during cooking?

4. How and why does excessive stirring affect palatability of cereal grains?

5. What are the advantages and disadvantages of cooking cereal products in the top of a double boiler, over boiling water?

6. How does instant rice differ from regular rice?

7. How is rice enriched? Should enriched rice be rinsed before cooking? Explain.

8. How can cereals be made more nutritious during the cooking process?

9. Complete the following table by reading labels or using appropriate references.

Product (1 c [240 ml] cooked)	Energy (kcal)	Protein (g)	Ca (%)	Iron (%)	Thiamin (mg)	Riboflavin (mg)
Rice, enriched						
Macaroni, enriched						
Grits, enriched						
Grits, unenriched						

10. Define the term *complex carbohydrate*.

11. Identify the contribution that cereals make to the fiber content of diets.

12. Describe proper storage of whole grains.

13. Identify the costs of various rice products:

Product	Cost/pkg	Package Size	Portion Cost
Long-cooking rice			
Instant rice			
Microwave rice			
Long-cooking rice, flavored			
Instant rice, flavored			
Microwave rice, flavored			

C. Fruits and Vegetables

OBJECTIVES

To know and describe parts of the parenchyma cell and how these are affected by heat
To relate cellular structure and principles of osmosis to recrisping vegetables
To evaluate the effect of processing (preparation, heat, water, alkali) on vitamin C retention in vegetables and fruit
To evaluate the effect of processing on the sugar content of vegetables
To relate nutrient losses to changes in the structure of a plant cell during processing
To describe the effects of pH on the color characteristics of raw fruits
To compare the effectiveness of various factors in preventing enzymatic browning in fruits
To evaluate the effects of cooking procedures on texture of cooked fruits
To know and describe the effect of pH changes on color and texture of vegetables
To evaluate the effect of cooking procedures on the color, flavor, and texture of vegetables
To illustrate palatable combinations of vegetables, noting contrasts in flavor, color, and texture
To demonstrate ability to apply principles of vegetable cookery by preparing various vegetable menu items
To appraise the nutritive, economic and sanitary dimensions of fruits and vegetables

REFERENCES

Appendices F, J, L, M, N

ASSIGNED READINGS

TERMS

Parenchyma	Tannin	Solute	*Brassica*
Cellulose	Anthoxanthin	Solvent	*Allium*
Cell sap	Chlorophyll	Solution	Boil
Vacuole	Pheophytin	Osmosis	Glaze
Plastid	Carotene	Diffusion	Bake
Nucleus	Steam	Permeable	Mince
Cytoplasm	Sauté	Semipermeable	Chop
Cell membrane	Stir Fry	Enzyme	Dice
Succulent	Pan Fry	Substrate	
Turgor	Panned	Enzymatic browning	
Anthocyanin	Pectic substances	Polyphenol	

FRUITS AND VEGETABLES

Courtesy: United Fresh Fruit and Vegetable Association

EXERCISE 1: PROPERTIES OF PARENCHYMA CELLS

A. COMPONENTS OF PARENCHYMA CELL

1. Label parts of the parenchyma cell.

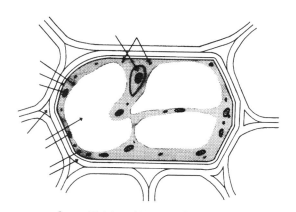

2. Identify fat- and water-soluble cell components.

3. List the major components of cell sap.

Source: Division of Nutritional Sciences, New York State College of Human Ecology at Cornell.

4. What is the effect on vacuole components of cutting through a cell (chopping, mashing)?

5. When living parenchyma cells are placed in water, how is the material in the vacuole affected? Why?

6. What changes in cell structure occur when a plant food is cooked?

7. When parenchyma cells are *cooked* in water, how is the material in the vacuole affected? What losses would you expect? Why?

8. Define and distinguish between osmosis and diffusion.

B. RECRISPING SUCCULENTS

PROCEDURE
1. For each treatment, process ½ c (120 ml) limp celery as directed.
2. Record observations as to crispness and turgor after holding 2 hours.

Treatment	Observations
1 c (240 ml) tap water	
1 c (240 ml) salt solution (3 T [45 ml] salt/1 c [240 L] water)	

1. Why is a *solution* of salt and water used, and not a mixture?

2. Based on these observations, what is the most effective method of recrisping vegetables?

3. Explain, using scientific principles and a diagram, the process of recrisping.

4. In vegetables that are recrisped for a long time would nutrient losses be expected? Explain.

5. List several vegetables that can be recrisped.

6. What storage methods keep vegetables crisp and eliminate the need for recrisping?

7. Based on these experiments, predict what would happen to raw peach slices sprinkled with sugar. Explain.

EXERCISE 2: ASSESSING NUTRIENT LOSS IN FRUITS AND VEGETABLES

A. Effect of Cutting, Chopping, and Soaking on Vitamin C (Ascorbic Acid)

PROCEDURE

Test for the presence of vitamin C[1] in uncooked fruit (orange, grapefruit, melon, apple, peach), broccoli and cabbage as follows:

1. Peel the food if necessary and finely chop. Conserve juice of fruits as much as possible.
2. Measure 2 T (30 ml) of the chopped food and juice and place it in a beaker or custard cup containing 30 ml water. Allow to soak 15 minutes, then drain, saving liquid. Use for vitamin C test.
3. Place 10 drops of 0.1% dichlorophenol indophenol[2] in a test tube. Initially the dye is dark blue.
4. Add the food solution, drop by drop to the dye. Shake the tube vigorously after the addition of each drop.
5. If the dye solution turns colorless, vitamin C is present. *Record* the number of drops of food solution used to turn the dye colorless. Do not use more than 80 drops of food solution. Note: The *more drops* of a food solution required to turn the dye colorless, the *less vitamin C* present in the solution.
6. Summarize the effect of cutting, chopping and soaking on vitamin C loss in uncooked fruits and vegetables.

---

[1] Appendix J
[2] To prepare stock solution, dissolve 13 mg 2,6-dichlorophenol indophenol in 100 ml triple-distilled water. Dilute solution 1:10 with triple-distilled water and filter. Refrigerate up to 5 days. Dye obtained from Fisher Scientific Co., Rochester, NY 14624.

Food	Number of Drops of Food Solution Used	Relative Amount of Vitamin C in Solution
Raw citrus fruit solution		
Raw fruit solution		
Raw cabbage solution		
Raw broccoli solution		

Conclusions:

B. Effect of Length of Cooking Time on Vitamin C

PROCEDURE

Test for the presence of vitamin C in the cooking water of either cabbage or broccoli as follows:

1. Place 2 T (30 ml) of chopped food in each of 2 small saucepans. Add 30 ml of water to each saucepan. Cook the food uncovered.
2. Cook one sample for 7 minutes, the other for 25 minutes. Start timing when mixture begins to boil.
3. As water will evaporate from the saucepans, add water during cooking to avoid burning and to maintain a *constant* 30 ml volume.
4. When each sample is finished cooking, drain vegetable and measure cooking liquid remaining in the saucepan. If 30 ml of liquid are not present, add water to make up the volume to 30 ml. *Constant volume is essential.* Cool cooking water.
5. Place 10 drops of dye in test tube. Test for ascorbic acid by adding cooking water, drop by drop. Shake vigorously after each drop. Record the number of drops of test solution needed to turn the dye colorless.
6. Summarize effects of length of cooking time on vitamin C loss.

FRUITS AND VEGETABLES

Food	Number of Drops of Food Solution Used	Relative Amount of Vitamin C in Solution
Broccoli water 7 minutes		
25 minutes		
Cabbage water 7 minutes		
25 minutes		

Conclusions:

C. Effect of Added Alkali on Vitamin C

PROCEDURE

Test for the presence of ascorbic acid in the cooking water of either broccoli or cabbage as follows:

1. Place 2 T (30 ml) of the chopped food and ¼ t (1.25 ml) baking soda (alkali) in a saucepan. Add 30 ml water. Cook uncovered for 7 minutes.
2. Follow steps 3 and 4 in Part B.
3. Test the cooking water for ascorbic acid as described in Part B. Record results below, in the chart.
4. Observe the effects of soda on color and texture of the vegetables.
5. Summarize effects of added alkali on vitamin C.

Food	Number of Drops of Food Solution Used	Relative Amount of Vitamin C in Solution
Broccoli, cooking water		
Cabbage, cooking water		

Conclusions:

1. Why was vitamin C detected in the soaking water?

2. Explain the difference in the amount of vitamin C found in cooking water and soaking water of the same vegetable.

3. Summarize the effect of soda on cooking a green vegetable.

	No soda	Soda	Explanation
Texture			
Vitamin C in cooking water			
Color of vegetable			
Color of cooking water			

4. Based on these experiments, what suggestions could you make as to the best way to prepare and cook a vegetable so it will have a high vitamin C content?

5. Based on these experiments, what precautions would you suggest when preparing fruits and vegetables for either canning or freezing?

EXERCISE 3: FRUITS

A. Enzymatic Browning

PROCEDURE

1. Prepare test solutions as indicated on the chart, prior to cutting fruit.
2. Test pH using Alkacid test paper.[1]
3. Peel assigned fruit (e.g., apples, bananas) and cut into 14 uniform slices.
4. For each treatment, use 2 fruit slices and thoroughly coat with test solutions. Leave 2 slices untreated.
5. Expose all fruit slices to air for 30 minutes. Record observations as to any color changes.
6. Summarize your conclusions as to the most effective way to prevent enzymatic browning.

Test Solution	pH	Color After Exposure to Air	Explanation
Untreated			
¼ c (60 ml) lemon juice			
¼ c (60 ml) pineapple juice			
¼ c (60 ml) water + 2 T (30 ml) sugar			
¼ c (60 ml) water + ¼ t (1.25 ml) cream of tartar			
¼ c (60 ml) water + ⅛ t (.63 ml) fruit fresh			
¼ c (60 ml) water + ⅛ t (.63 ml) ascorbic acid)			

[1]Test paper obtained from Fisher Scientific Co., Rochester, NY 14624.

1. What components must be present for a fresh fruit or vegetable to brown?

2. What additives are used commercially to prevent browning of fresh/dried fruits? Potatoes?

B. Effect of Sugar on Texture and Flavor of Cooked Fresh Fruit

Procedure

1. Prepare sugar solution as directed.
2. Place each solution in a small, shallow pan.
3. Peel, core and cut 2 cooking apples to make 16 uniform pieces.
4. Place 4 slices in each pan and cook *covered* for 6 to 8 minutes over medium heat. (Slices must be submerged in solution.)
5. Record observations noting the movement of water and/or solutes that takes place during cooking.
6. Use palatability terms below and account for palatability characteristics that were observed.

Test Solution	Osmosis	Diffusion	Palatability	Explanation
½ c (120 ml) water				
½ c (120 ml) water + 2 T (30 ml) sugar				
½ c (120 ml) water + 1 c (240 ml) sugar				
½ c (120 ml) water—cook 6 min; + 1 c (240 ml) sugar—cook 2 more min				

```
                    PALATABILITY TERMS
    Texture         Flavor                  Shape
    Firm            Uniform sweetness       Retains shape
    Soft            Surface sweetness       Sauce
    Mushy           Retains natural flavor  Shrunken
    Glazed          Bland
```

1. Under what conditions does osmosis cease?

2. Are nutrients lost from fruit during osmosis? Explain.

3. Are nutrients lost from fruit during diffusion? Explain.

4. What are the effects of large amounts of sugar on plant pectins?

C. Effect of Sugar on Texture and Flavor of Cooked Dried Fruit

Procedure

1. Cook dried fruits in boiling test solution for 10 minutes.
2. Record observations and account for palatability.

Test Solution	Palatability	Explanation
¼ c (60 ml) raisins + ⅓ c (80 ml) water[1]		
¼ c (60 ml) raisins + ⅓ c (80 ml) water + ¼ c (60 ml) sugar		

[1]Check to maintain sufficient test solution in saucepan.

Conclusions:

1. How does drying affect plant membranes?

2. What process occurs during the cooking of dried fruits?

D. Factors Affecting Anthocyanin Pigments

PROCEDURES

1. Mix 2 T (30 ml) assigned fruit juice (e.g., grape, blackberry, or cranberry) with each variable.
2. Record and explain results on chart.
3. Summarize conclusions.

Variable	Color	Explanation
1 T (15 ml) lemon juice		
2 T (30 ml) orange juice		
2 T (30 ml) pineapple juice		
2 T (30 ml) strong tea		
¼ t (1.25 ml) baking soda		
Iron chloride (few drops)		

Conclusions:

1. Is enzymatic browning a problem in the following? Explain.

   canned apple slices:

   pear slices wrapped in plastic wrap:

   fresh fruit cup (grapes, cheese cubes, bananas):

2. Why do some fruits (apples, pears, bananas) turn brown when sliced or bruised, but the intact fruit does not brown?

3. To achieve a palatable cooked product (soft, sweet, tender) from a mixture of dried fruits, how should they be prepared? Why?

4. Provide examples of how one would apply information about reaction of anthocyanin pigments when working with whole fruits or juice mixtures.

5. What are the major nutrient contributions of fruits? Specify which fruits are *not* excellent sources of vitamin C.

6. Account for the different values for fiber in the three apple products shown. What components of the whole apple contribute fiber?

Whole apple with peel: 3.6 g fiber   Applesauce, 1/2 cup: 2.1 g fiber   Apple juice, 3/4 cup: 0.2 g fiber

7. Complete the following nutritive value chart:

	Energy (kcal)	Iron (mg)	Vitamin A (IU) or (RE)	Vitamin C (mg)	Fiber (g)
Apple (1 medium)					
Apricots, dried (½ c, 120 ml)					
Cantaloupe (¼)					
Grapefruit, pink (½)					
Grapefruit, white (½)					
Orange (1 medium)					
Pear (1 medium)					
Peach (1 medium)					
Peaches, canned (½ c; 120 ml)					
Pineapple, canned (2 sl)					
Prunes, dried (½ c; 120 ml)					
Prunes, stewed (½ c; 120 ml)					
Strawberries (½ c; 120 ml)					

8. Explain how color can give an indication of the relative amounts of vitamin A in a fruit.

Courtesy: Florida Citrus Commission

COOKING VEGETABLES—PIGMENTS, PALATABILITY, PRECAUTIONS

PIGMENTS

GREEN (chlorophyll)
  beet greens
  broccoli
  butter beans (lima beans)
  collards
  green beans
  green cabbage
  kale
  peas
  spinach
  swiss chard

YELLOW (carotene)
  carrots
  rutabagas
  summer squash
  sweet potatoes
  wax beans
  winter squash (hubbard, butternut)
  yams
  yellow turnips

RED (anthocyanin)
  beets
  red cabbage

WHITE (anthoxanthin)
  cauliflower
  onions
  white potato
  white turnip

FLAVOR

Allium spp.
  chive
  garlic
  leeks
  onion
  shallots

Brassica spp.
  broccoli
  Brussels sprouts
  cabbage
  cauliflower
  kale
  turnip
  rutabagas

PALATABILITY TERMS—VEGETABLES		
TEXTURE	COLOR	FLAVOR
Hard	Natural	Natural
Firm	Red-Blue	Bland
Crisp	Blue-green	Mild
Tender	Gray-blue	Strong
Soft	Pink	Off-flavor
Mushy	Green–bright green, olive green	
	White–cream	
	Yellow	

As a precaution against contamination and growth of harmful bacteria[1]:

- Carefully wash hands *before* handling fresh produce, and *between* cutting and eating fresh produce.
- Rinse produce to remove harmful bacteria prior to consumption (even uneaten, disposed of rinds), and after removing outer leaves and peels.
- Prepare fruits and vegetables on sanitary work surfaces with sanitary utensils.
- When fruits and vegetables are cut, inadvertent contamination has occurred; therefore, cover and refrigerate during storage to slow any bacterial growth.

[1] Adapted from Tufts University *Diet & Nutrition Letter* 14(11):1–2, 1997.

EXERCISE 4: COOKING VEGETABLES[1]

A. Effect of pH on Pigments and Texture

PROCEDURE

1. Select vegetables characteristically colored by the pigments, as assigned. (Spinach, carrots, red cabbage, white potato.)
2. For each vegetable, peel (if necessary) and cut into uniform size pieces.
3. For each treatment place 1 c (240 ml) assigned vegetable and ½ c (120 ml) boiling water in a small saucepan. Add soda or vinegar as assigned.
4. Measure pH of water using *Alkacid* test paper and record.
5. Cover pan and bring water back to a boil. Start timing the cooking.[2]
6. After 10 minutes, remove ½ of the vegetable. Label and display. Cook the remaining portion 15 more minutes. Drain, label, and display.

[1] All fruits and vegetables intended for consumption should be washed well before preparation and cooking.
[2] Check to maintain sufficient water in saucepan.

7. Compare color and texture of all samples. Record observations.

Vegetable Treatment	pH	Cooked 10 minutes		Cooked 25 minutes	
		Color	Texture	Color	Texture
Chlorophyll water only (control)					
+ ½ t (2.5 ml) soda					
+ 2 T (30 ml) vinegar					
Carotene water only					
+ ½ t (2.5 ml) soda					
+ 2 T (30 ml) vinegar					
Anthocyanin water only					
+ ½ t (2.5 ml) soda					
+ 2 T (30 ml) vinegar					
Anthoxanthin water only					
+ ½ t (2.5 ml) soda					
+ 2 T (30 ml) vinegar					

B. Effect of Cooking Procedure on Pigments and Flavors

PROCEDURE

1. Select vegetables characteristically colored by chlorophyll and anthocyanin pigments.
2. Select vegetables from the *Allium* and *Brassica* families.
3. Prepare 1 c (240 ml) vegetable for each variable. Follow directions in table for *use of cover and amount of water*.
4. Add water; start timing after water returns to a boil.
5. Remove ½ of the vegetable after cooking 10 minutes. Label and display. Cook remaining portion 15 more minutes. Label and display.
6. Evaluate all vegetables, and record observations.

Pigment	Cover	Amount of water	Cooked 10 min		Cooked 25 min	
			Color	Texture	Color	Texture
Chlorophyll 1.	On	½ c (120 ml)[1]				
2.	Off	to cover				
Anthocyanin 1.	On	½ c (120 ml)				
2.	Off	to cover				
Flavor			Color	Flavor	Color	Flavor
Allium 1.	On	½ c (120 ml)[1]				
2.	Off	to cover				
Brassica 1.	On	½ c (120 ml)				
2.	Off	to cover				

[1] Check to maintain sufficient water in saucepan.

1. Summarize the effect of *long cooking time* on factors listed below:

	Effect of Time	Explanation
A. Texture		
B. Color Chlorophyll		
Carotene		
Anthocyanin		
Anthoxanthin		
C. Flavor *Allium* (onion family)		
Brassica (cabbage family)		
D. Nutritive Value Vitamin A		
Vitamin C		
Thiamin		

2. Summarize the effect of covering the pan when cooking the following:

Vegetable	Effect of Cover	Explanation
Green pigment		
Red pigment		
Allium flavor		
Brassica flavor		

3. Referring to the experiment in Part B:
   a. Why was there no variable in which 1 c (240 ml) vegetable was cooked in ½ c (120 ml) water, uncovered?

   b. Predict the effect on flavor of a vegetable from the *Allium* and *Brassica* family if they had been cooked, covered with water, in a pan with a cover.

C. APPLICATION OF PRINCIPLES TO COOKING A VARIETY OF VEGETABLES

PROCEDURE

1. Examine the display of raw vegetables, observing characteristics that denote freshness and excellent quality. Consider percentage waste as vegetables are prepared.
2. Prepare assigned vegetable recipe and plan cooking times (item with longest estimated preparation and cooking time should be first) to serve the product in _____ hour(s). If assigned recipe is to be cooked in the microwave, refer to the vegetable section of the microwave chapter.
3. Evaluate the palatability of all products, and complete chart on pigment, flavor, and palatability.
4. Calculate the nutritive value of one recipe.

Evaluation of Vegetable Recipes

Vegetable Recipe	Major Pigment	Major Flavor	Palatability

Evaluation of Vegetable Recipes

Vegetable Recipe	Major Pigment	Major Flavor	Palatability

NUTRITIVE VALUE OF RECIPES

Food	Measure	Energy (kcal)	Protein (g)	Calcium (mg)	Iron (mg)	Total Vit A Act. (IU) or (RE)	Vitamin C (mg)	Thiamin (mg)	Riboflavin (mg)

VEGETABLE RECIPES[1]

BEETS

Panned Beets

1. Peel beets (1½ c; 360 ml) and remove the stem. Slice, dice or shred the beets.
2. Heat 1 T (15 ml) oil in a heavy skillet or saucepan.
3. Add beets and toss until the vegetable is coated with oil. Add a small amount of water if necessary. Turn the heat down and stir to prevent burning. Vegetable should be crisp in texture. (2 servings)

BROCCOLI

1. Remove coarse leaves and tough parts of stalk.
2. Split stalks, peel if tough; leave 3-inch (7.5-cm) stem on flowerets.
3. Boil uncovered in a small amount of water for 3 minutes. Cover, cook 5 to 8 minutes longer until just tender. Drain.
4. If desired, add 1 t (5 ml) margarine to each cup (240 ml) cooked vegetable; mix lightly. Season.

Steamed Broccoli Medley

bay leaf	1	1	cauliflower pieces	1 c	240 ml
broccoli flowerets	1 c	240 ml	snow peas	½ c	120 ml
carrot strips	½ c	120 ml			

1. Pour 1 inch (2.54 cm) water and bay leaf into a medium saucepan. Boil.
2. Place all vegetables into expandable steamer basket.
3. Insert basket into pot and cover with a tight-fitting lid.
4. Steam the vegetables for approximately 8 to 10 minutes, or until vegetables are tender crisp.
5. Season. (4 to 6 servings)

CABBAGE

Green Cabbage

Remove wilted outside leaves, wash carefully. Cut in wedges or shred. Cook uncovered in water to cover. (Wedges, 10 to 12 min.; shredded, 5 to 8 minutes)

Cabbage with Tomato Sauce

green cabbage, shredded	1½ c	360 ml	tomato sauce	¾ c	180 ml
onion, minced	¼ c	60 ml	brown sugar	1 t	5 ml
slice bacon, diced (optional)	1	1			

1. Cook cabbage, uncovered, in water to cover, about 7 minutes. Drain.
2. Sauté onion with bacon (or use 1 t [5 ml] oil) until tender.
3. Add tomato sauce and sugar to onion.
4. When sauce comes to a boil, add well drained cabbage. Season. (2 servings)

[1] See Microwave Cooking chapter for microwave recipes. Use seasonings to taste.

Fruits and Vegetables

Pennsylvania Red Cabbage

vegetable oil	½ T	7.5 ml	water		1 T	15 ml
red cabbage, shredded	1½ c	360 ml	caraway seed		⅛ t	.63 ml
unpared apple, cubed	½ c	120 ml	vinegar		½ T	7.5 ml
brown sugar	1 T	15 ml				

1. Heat oil in skillet; add remaining ingredients, except vinegar.
2. Cover tightly; cook over low heat, stirring occasionally.
3. Cook 15 to 25 minutes until desired tenderness is reached.
4. Stir in vinegar. Season. (2 to 3 servings)

CARROTS

Scrub. Pare if desired. Leave whole, or cut into crosswise or lengthwise slices. Boil in a small amount of water, covered. (10 to 20 minutes).

Glazed Carrots

medium carrots	3	3	brown sugar	1½ T	22.5 ml
margarine	1½ T	22.5 ml			

1. Cook carrots covered in small amount of boiling water for about 6 to 10 minutes or until tender. Drain.
2. In a skillet, melt fat and add brown sugar. Stir until melted.
3. Add the cooked carrots and cook slowly, stirring until the carrots are well glazed. Season. (2 servings)

CAULIFLOWER

1. Remove leaves and some of the woody stem from ¼ head cauliflower. Separate into flowerets.
2. Cook covered in a small amount of water 10 to 15 minutes or until just tender.
3. Drain and, if desired, add 1 t (5 ml) margarine per cup (240 ml) cooked vegetable. Season.

Greek Cauliflower

vegetable oil	1 T	15 ml	canned tomatoes	⅔ c	160 ml
finely chopped onion	2 T	30 ml	lemon juice	2 T	30 ml
cauliflower, pieces	2 c	480 ml	basil	½ t	2.5 ml
water	¼ c	60 ml			

1. In a medium skillet, heat oil. Add onion and sauté until tender.
2. Add cauliflower and water. Cover and cook until just tender.
3. Add chopped tomatoes, lemon juice, and basil to cauliflower.
4. Bring to boil. Reduce heat, cover and simmer 4 to 6 minutes until flavors blend. Season. (3 to 4 servings)

COLLARDS

Sautéed Collard Greens

bacon	1 slice		vegetable oil	1½ T	45 ml
collard greens, fresh	½ lb	227g	garlic	1 clove	

1. Fry bacon, drain. Crumble and reserve.
2. Blanch shredded greens for 3 minutes. Drain.
3. In a medium fry pan heat oil with garlic for a few minutes. Remove garlic. Add the blanched greens and water. Stir.
4. Cover the pan and cook the greens over low heat for 15 minutes, stirring occasionally.
5. Drain if necessary. Sprinkle with bacon bits. (2–3 servings)

EGGPLANT

Baked Eggplant

small eggplant	1	1	dry bread crumps	½ c	120 ml
evaporated milk	⅓ c	80 ml	seasonings		

1. Set oven at 400°F (205°C).
2. Peel eggplant and slice ¼ inch (.63 cm) thick.
3. Dip eggplant into evaporated milk.
4. Combine crumbs and seasonings. Dip eggplant into crumb mixture.
5. Bake on baking sheet 10 to 12 minutes or until tender. Serve with Cheese Sauce. (2 to 3 servings)

Cheese Sauce

processed cheese	¼ c	60 ml	evaporated milk	⅓ c	80 ml

Combine milk and cheese in heavy saucepan or top of double boiler and cook until cheese melts. Stir frequently. If too thick, thin with milk or evaporated milk. Pour over baked eggplant.

Ratatouille (Eggplant–Vegetable Stew)[1]

chopped onion	¼ c	60 ml	sliced zucchini	1½ c	360 ml
chopped green pepper	½ c	120 ml	canned tomatoes	¾ c	180 ml
vegetable oil	2 T	30 ml	tomato sauce	2 T	30 ml
diced eggplant	2 c	480 ml	basil, oregano	¼ t	1.25 ml

1. In a heavy saucepan or skillet, sauté onion and green pepper in oil until soft.
2. Stir in eggplant and zucchini. Sauté 5 minutes, adding a little more oil, if needed, to prevent sticking.
3. Add tomato, tomato sauce, and seasonings.
4. Cover and boil gently about 25 minutes or until vegetables are tender. Season. (4 to 5 servings)

[1] Other vegetables, such as potato, summer squash, mushrooms, and celery, make flavorsome additions.

FRUITS AND VEGETABLES

GREEN BEANS

Wash, remove ends, leave whole or cut crosswise or lengthwise. Boil, uncovered, in small amount of water for a few minutes. Cover and cook until tender. (Whole, 15 to 20 minutes; cut, 8 to 12 minutes.)

Green Beans with Mushrooms
(Habichuelas con Hongos)

green beans, sliced	2 c	480 ml	red pimentos, cut into strips	2	2
olive oil	1 T	15 ml	cooked mushrooms, sliced	½ c	120 ml
onion, minced	2 T	30 ml			

1. Cook green beans, covered, in small amount of water until barely tender.
2. Sauté onion in hot oil. Add cooked beans and pimento. Sauté together for about 5 minutes. Add mushrooms. Heat thoroughly, stirring gently. Season. (3 to 4 servings)

ONIONS

Peel under running water. Leave whole, slice or quarter. Boil in a large amount of water, uncovered (Slices: 10 min; whole 35 min.)

Cheese-Scalloped Onions

onion, sliced	1½ C	360 ml	milk	½ c	120 ml
margarine	1½ T	22.5 ml	processed cheese	¼ c	60 ml
all purpose flour	1½ T	22.5 ml			

1. Set oven at 350°F (175°C).
2. Cook onions uncovered in a large amount of boiling water until nearly tender (6 to 8 minutes); drain well. Place drained onions in a small greased casserole.
3. Melt fat in saucepan, blend in flour.
4. Add milk, cook while stirring until mixture boils.
5. Stir in cheese. Pour sauce over onions.
6. Bake casserole, uncovered about 10 to 15 minutes. Season. (4 servings)

PLANTAINS

Fried Green Plantains
(Tostones)

1. Peel 2 large green plantains and cut diagonally into slices ½ inch (1.25 cm) thick.
2. Soak slices in a salt solution for about 10 minutes.
3. Dry the slices and fry in vegetable oil at medium heat for about 10 minutes.
4. Remove slices, place on absorbent paper. Fold the paper in half over the slices, and press hard until the slices have been flattened.
5. Refry until golden brown, for about 5 minutes.
6. Remove and drain on absorbent paper to absorb excess oil. Season. Fried plantains may be made ahead and reheated in 400°F (205°C) oven for 5 to 10 minutes. (Note: the recipe may be high in salt.)

POTATOES

Baked Potatoes

1. Set oven at 425°F (220°C).
2. Select smooth potatoes of uniform size. Scrub thoroughly. If soft skin is desired, rub with oil before baking. Prick skin with fork to allow steam to escape.
3. Bake 45 to 60 minutes or until soft.
4. Remove from oven. Serve promptly. If not served immediately, soften potato by rolling in hands, protected with a clean towel.

Stuffed Baked Potatoes

1. Set oven at 425°F (220°C).
2. Cut a slice from top of baked potato or cut in half lengthwise.
3. Scoop out potato pulp, being careful not to break the skin.
4. Mash and season potatoes. (Use 1 to 2 T [15 to 30 ml] milk, 1 t [5 ml] margarine.)
5. Pile mixture lightly in the skins, leaving top rough.
6. Place in pan and bake about 10 minutes until delicately browned. Garnish with finely chopped parsley, grated cheese, or paprika.

Stuffed Yams

medium sweet potato or yam	1	1	milk	1–2 T	15–30 ml
margarine	1 T	15 ml	chopped walnuts	1 T	15 ml
brown sugar	½ t	2.5 ml			

1. Set oven at 400°F (205°C).
2. Scrub potato. Bake 40 minutes or until potato tests done with a fork.
3. Cut potato in half. Scoop out inside being careful not to break shell.
4. Mash potatoes in a mixing bowl. Add remaining ingredients, except nuts, with enough hot milk to moisten.
5. Beat until fluffy. Fold in nuts. Pile mixture back into potato shell.
6. Bake 15 minutes or until heated through. Season. (1 serving)

SPINACH

Remove root ends and damaged leaves. Break off large stems. If necessary, wash several times until leaves are free of grit and sand. Use only water that clings to the leaves, or a small amount of water. Cook 5 to 7 minutes in a covered pan.

Stir-Fried Spinach

peanut oil	1½ T	22.5 ml	spinach	½ lb	227 g
clove garlic, crushed	1	1	water chestnuts	¼ c	60 ml

1. In a medium frying pan, heat oil with garlic over high heat for a few minutes. Remove garlic.
2. Add spinach and cook, stirring gently approximately 2 minutes until spinach is heated through. Add water chestnuts and season. (2 servings)

TOMATOES

Baked Tomato

margarine	1 T	15 ml	prepared mustard	½ t	2.5 ml
onion, finely chopped	2 T	30 ml	Worcestershire sauce	¼ t	1.25 ml
soft bread crumbs	¼ c	60 ml	tomatoes	2	2

1. Set oven at 400°F (205°C).
2. Melt margarine. Add onion and crumbs; sauté until onion is soft. Add seasonings.
3. Halve tomato crosswise. Scoop out inside and add to crumb mixture.
4. Fill tomato with mixture. Bake 25 to 30 minutes. Season. (2 servings)

Mexican Succotash

margarine	1 T	15 ml	corn (fresh, canned or frozen)	½ c	120 ml
onion, chopped	1 T	15 ml			
zucchini, sliced	1 c	240 ml	canned tomatoes	¼ c	60 ml

1. Sauté onion in fat until soft.
2. Add zucchini and corn. Cook covered, 15 minutes or until tender.
3. Add tomatoes; cook just to heat through. Season. (2 servings)

ZUCCHINI

Zucchini Sauté

1. Wash 1 medium zucchini. Do not pare. Slice thin.
2. Cook, covered in 1 T (15 ml) margarine in skillet for 5 minutes.
3. Uncover and cook, turning slices until just tender. Season to taste. Sprinkle with 1 T (15 ml) parmesan cheese, if desired. (2 to 3 servings)

Zucchini Cheese Casserole

margarine	2 T	30 ml	oregano	¼ t	1.25 ml
chopped onion	¼ c	60 ml	canned tomatoes	½ c	120 ml
sliced zucchini	2 c	480 ml	processed cheese	¼ c	60 ml

1. Melt fat in medium frying pan. Add onions, and sauté until tender.
2. Add zucchini and oregano, and cook covered, for 7 minutes or until zucchini is just tender.
3. Add tomatoes, cover and simmer 5 minutes longer. Sprinkle with cheese before serving. Season. (4 servings)

OTHER VEGETABLE MIXTURES
Stir-Fried Vegetables[1]

oil	2–3 T	30–45 ml	zucchini, sliced	½ c	120 ml
clove garlic, crushed	1	1	water chestnuts, sliced	¼ c	60 ml
broccoli, flowerets	1 c	240 ml	green pepper, sliced	½ c	120 ml
carrots, sliced	½ c	120 ml	snowpeas	½ c	120 ml
green beans, sliced	½ c	120 ml	mung bean sprouts	¼ c	60 ml

1. Cut vegetables to a similar small size or slice thin.
2. Heat oil in a large skillet or wok, until it is hot, but not smoking. Add garlic.
3. Add broccoli and carrots. Cook 1½ minutes, lifting and turning to expose all sides of the food to the hot pan surface.
4. Add green beans and cook 1 minute.
5. Add zucchini and water chestnuts and cook 1½ minutes. Cover and let mixture steam 1 to 2 minutes, adding 2 to 3 T (30 to 45 ml) water if dry.
6. Add green pepper, snow peas and sprouts and cook 1 minute. (4 to 6 servings)

Fat must be extremely hot, but not smoking for *each* addition of a new vegetable. Vegetables must be stirred as they are cooked. Vegetables will be *done* when opaqueness disappears and translucency begins to appear. Vegetables should be crisp, but not taste raw, and hot.

STIR FRYING

Courtesy: United Fresh Fruit and Vegetable Association

1. Predict the effect on color when milk (pH 6.6) is added to mashed potatoes.

2. Why are Cheese-Scalloped onions especially mild in flavor?

---

[1] Vegetables may be varied as to type and amount used. Preparation of the vegetables will take the most time, but cooking time is short.

3. Account for differences in green color observed between Zucchini Sauté and Zucchini Casserole.

4. Referring to the glazed carrot recipe, why are the carrots cooked until tender *before* the sugar is added?

5. In Greek Cauliflower, why are the tomatoes added *after* the cauliflower is tender?

SUMMARY QUESTIONS—FRUITS AND VEGETABLES

1. Which vegetables are excellent sources of vitamin A? Vitamin C?

2. As a group, what major nutrients do fruits and vegetable contribute?

3. In addition to vitamins and minerals, what other value do fruits and vegetables have in the daily diet?

4. In general which nutrients are most likely to be lost or destroyed during cooking? Indicate what processing factors cause loss.

5. Based on laboratory experiments, what general principles can be followed to *minimize* nutrient loss and *enhance* palatability?

6. Complete the following table with appropriate information:

	Pigments	Flavors	Nutrients
Water soluble			
Fat soluble			
Volatile			

7. Discuss potential sanitary problems associated with fruits and vegetables. How may these problems be controlled?

8. What storage conditions for vegetables and fruits should be maintained in a grocery store? Why?

9. Provide brief directions for storage of fruits and vegetable in the home.

10. Compare the cost of three fruits and/or vegetables in different forms:

COST COMPARISON

Fruit/Vegetable	Fresh	Frozen	Dried	Canned
1.				
2.				
3.				

Which is the best buy? How would this change with the season? How would this change with intended use?

11. Summarize specific factors that affect cost of fruits and vegetables in the marketplace.

D. Meat, Poultry, and Fish

OBJECTIVES

To recognize common retail and primal cuts of meat
To relate location of cut and species to inherent palatability characteristics
To differentiate effects of dry and moist heat on meat products
To demonstrate ability to apply principles of preparation to meat, poultry, and fish
To know and apply principles of sanitary quality to preparation of meat, poultry, and fish
To appraise nutritive and economic dimensions of meat, poultry and fish products

REFERENCES

Appendices G-I, G-II, N

ASSIGNED READINGS

TERMS

Myosin	Moist heat	Broil	Pressure cook
Collagen	Dry heat	Pan broil	Cure
Elastin	*Escherichia coli*	Bread	Tenderize
USDA	*Salmonella*	Stew	Inherent tenderness
Marbling	*Clostridium perfringens*	Braise	Stock
Grain	Trichinosis	Poach	Gelatin
Conformation	Roast	Baste	Steak
Wholesomeness	Bake	Pot roast	Fillets

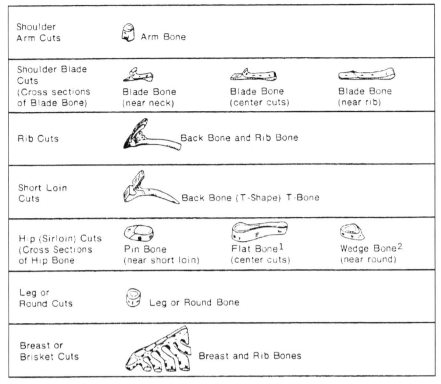

BONES IDENTIFYING SEVERAL GROUPS OF RETAIL CUTS

Source: USDA, National Cattlemen's Beef Association

[1] Formerly part of "double bone," but today the back bone is usually removed, leaving only the "flat bone" (sometimes called "pin bone") in the sirloin steak.

[2] On one side of sirloin steak, this bone may be wedge shaped, while on the other side, the same bone may be round.

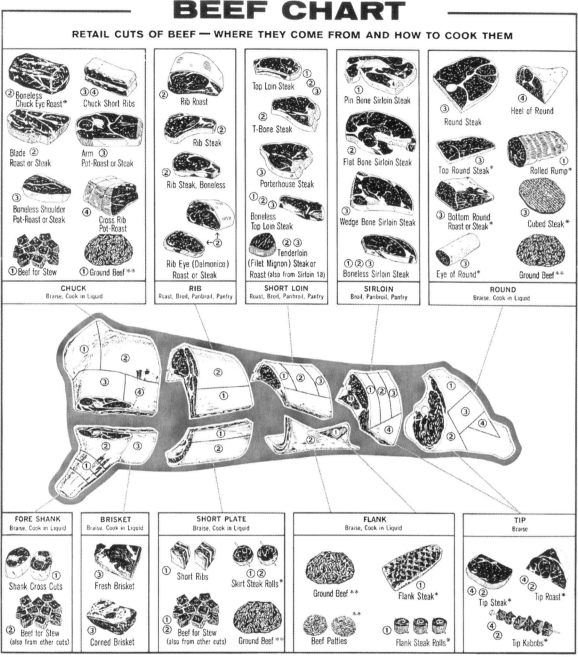

BEEF CHART

Courtesy: USDA, National Cattlemen's Beef Association

EXERCISE 1: IDENTIFICATION OF BASIC MEAT CUTS

PROCEDURE

1. Study the cuts of meat on display or visit a meat market where various cuts of meat are on display
2. Complete the following table:

Name of Cut	Primal Cut	Inherent Tenderness	Characteristic Bone, Color, Grain Marbling	Cost/ lb (454 g)	Cost/ Serving
Beef Blade steak					
Arm steak					
Rib steak					
T-bone or sirloin steak					
Flank steak					
Bottom round					
Top round					
Pork Loin chop					
Rib chop					
Veal Blade steak					
Cutlet					
Lamb Rib chop					
Blade chop					

MEAT, POULTRY, AND FISH

1. Identify the following markings found on meat and meat products. What do they indicate regarding meat quality and wholesomeness?

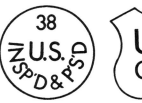  

LABELS ON MEAT PRODUCTS
Courtesy: USDA

2. What government agency is responsible for inspecting meat and poultry products, both fresh and processed?

3. What generalizations can be made regarding location of a cut of meat and its inherent tenderness?

4. Which is the best way to calculate cost of meat; price per pound or price per serving? Explain.

5. Indicate the approximate number of servings per pound (454 g) of the following products:

   beef round: liver: spare ribs
   sirloin: pork chops: ground chuck:

EXERCISE 2: EFFECT OF DRY AND MOIST HEAT ON LESS TENDER (TOUGH) CUTS OF MEAT

A. ROASTS

PROCEDURE

Observe a demonstration prepared as follows:

1. Weigh 2 roasts (e.g., chuck, bottom round) approximately 2½ lb (1.14 kg) each (paired cuts if possible). Insert meat thermometer into each as shown in the photograph.
2. Place one roast uncovered (Roast A) and one roast covered (Roast B) in a 325°F (165°C) oven.
3. Cook roasts until internal temperature of Roast A reaches 77°C (170°F). Record internal temperature of Roast B.
4. Remove both roasts from the oven, weigh and transfer to serving dish. Record total cooking time.
5. After cooling 10 minutes, slice and evaluate palatability using palatability terms. Summarize results.

112   Food Principles

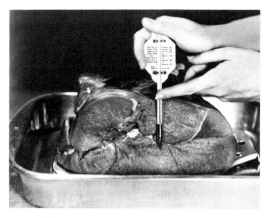

Demonstrating depth of meat thermometer
Courtesy: USDA

PALATABILITY TERMS	
MOISTNESS	**TENDERNESS**
Juicy	Crumbly
Moderately juicy	Tender
Moderately dry	Moderately tender
Dry	Moderately tough
	Tough

FLAVOR
Flavorful
Moderately flavorful
Moderately bland
Bland
Tasteless, no flavor

	Roast A—Dry Heat uncovered	Roast B—Moist Heat covered
Retail cut		
Primal cut		
Inherent tenderness		
Oven temperature	325°F (165°C)	325°F (165°C)
Length of cooking		
Final internal temperature	170°F (77°C)	
Weight loss		
Effect of cooking on: Muscle fiber		
Connective tissue		
Moistness		
Tenderness		
Flavor		

1. Relate the observed rate of heat penetration by the dry and moist methods of cooking.

2. Were the roasts equally tender and juicy? Explain.

3. Roast A (dry heat) was removed from the oven when the internal temperature reached 170°F (77°C). If the roast was allowed to cook longer, would the internal temperature increase? If so, how would this influence juiciness?

4. In cooking meat, what is the highest internal temperature obtainable by the moist heat method?

B. Meat Patties[1]

PROCEDURE

1. Prepare two 3-ounce (85-g) patties of ground chuck approximately ½ inch (1.25 cm) thick. Weigh.
2. Pan broil one patty, braise the other, following directions specified in the table below.
3. Weigh each patty after cooking and calculate cooking losses.
4. Evaluate both patties, assessing tenderness, juiciness, and flavor.

PAN BROIL	BRAISE
1. Place meat in hot, greased fry pan and cook *slowly* over moderate heat on one side for 7 minutes. 2. Pour off excess fat as it accumulates. 3. Turn and cook meat for an additional 7 minutes.	1. Brown meat in 1 t (5 ml) oil. 2. Add ⅓ c (80 ml) hot water to skillet. 3. Cover and simmer for 14 minutes.

---

[1] If internal temperature of meat patty has not reached 155°F (68°C), do not taste, but observe characteristics of tenderness and juiciness.

	Pan Broiled	Braised
Weight before cooking		
Cooked weight		
Cooking losses		
Tenderness		
Juiciness		
Flavor		

1. How are pan broiling and braising cooking methods classified? List other examples of the basic methods of cooking meat.

2. Applying principles of protein coagulation, explain the differences in juiciness obtained by the two methods of cooking.

3. What is the effect of grinding (ground beef) on muscle protein? On connective tissue?

4. What techniques, other than grinding, are used to "tenderize" meat? Explain how other techniques work.

EXERCISE 3: APPLICATION OF PRINCIPLES TO MEAT, POULTRY, AND FISH COOKERY

PROCEDURE

1. Inspect a display of meat, poultry, and fish products. Note species differences in texture, grain, color, and degree of marbling. Consider appearance as related to evidence of freshness.
2. Using assigned recipe, prepare product. Plan cooking so that the product will be be hot and ready to serve (   ) hours after the start of class.
3. Evaluate all prepared products.

Evaluation of Meat, Poultry, and Fish Recipes

Food	Appearance Uncooked	Inherent Tenderness	Method of Cooking	Sensory Evaluation	
				Tenderness	Juiciness
Variety meats					
Sweetbreads					
Liver—chicken					
Liver—					
Liver—					
Shellfish					
Shrimp					
Scallops					
Fish					
Poultry					
Chicken					
Other meats					
Veal					
Lamb					
Beef, steak					
Steak					
Steak, flank					
Pork chop					

116 FOOD PRINCIPLES

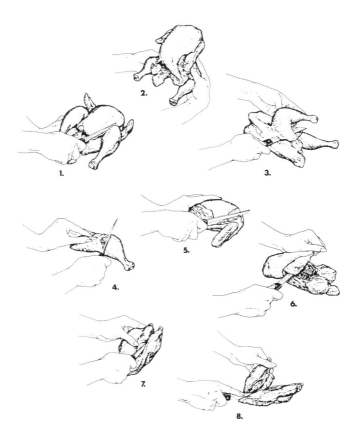

Cutting Up A Whole Chicken

Step 1. Place chicken on back on cutting board. Using sharp knife, cut skin, between thighs and body.
Step 2. Grasping one leg in each hand, lift chicken and bend back legs until bones break at hip joints.
Step 3. Turn bird on side. Remove leg and thigh from body by cutting from tail and toward shoulder. Cut between joints close to bones in back of bird. Repeat other side.
Step 4. Separate thighs and drumsticks. Locate knee joint by bending thigh and leg together. Cut through joints of each leg.
Step 5. With chicken on back, remove wings by cutting inside of wing over joint. Cut from top down, through joint.
Step 6. Separate breast and back by placing bird on neck end or back, and cutting through joints along each side of rib cage. Cut away from you toward board.
Step 7. Whole breast is ready to be used as is or you may bone it if desired.
Step 8. To split breast into halves, cut wishbone in two at V of bone. Halves may also be boned.

Courtesy: National Broiler Council

Courtesy: SYSCO ® Incorporated

Courtesy: SYSCO ® Incorporated

MEAT, POULTRY, AND FISH RECIPES[1,2]

VARIETY MEAT RECIPES

Sweetbreads

1. Soak sweetbreads for 1 hour in large quantity of cold water to release any blood.
2. Place in water containing 1 t (5 ml) salt and 1 T (15 ml) vinegar or lemon juice per quart (or liter for laboratory purposes).
3. Bring slowly to a boil, then simmer for 5 minutes.
4. Plunge into cold water to firm.
5. After sweetbreads have cooled, drain and trim them by removing cartilage tubes, connective tissue and tougher membranes.
6. Break sweetbreads into smaller sections, being careful not to disturb the fine membrane surrounding the smaller units.
7. After the preliminary preparation, sweetbreads may be poached, braised, broiled, fried, or served in a cream sauce.

Sautéed Sweetbreads

1. Dip cooked sweetbreads in melted margarine.
2. Pan-fry until brown, about 10 minutes. Season.
3. May be served plain or in a thin sauce (mushroom, or sweet and sour).

LIVER
Chopped Chicken Livers

egg	1	1	oil	1 T	15 ml
chicken livers	⅓ lb	151 g	mayonnaise	2 T	30 ml
finely chopped onion	¼ c	60 ml			

1. Simmer egg until hard. Cool, shell.
2. Simmer chicken livers until just tender. Drain and cool.
3. Sauté onion in oil. Add to liver and egg. Chop and mix until livers are a fine paste or force through a food grinder. Add mayonnaise, mix and season. Chill and serve with crackers.

Crisp Liver Strips

beef liver	⅓ lb	151 g	egg, slightly beaten	1	1
fat free French or Italian dressing	¼ c	60 ml	cracker crumbs	½ c	120 ml
			vegetable oil	2–3 T	30–45 ml

1. Cut liver into ½-inch (1.25-cm) strips.
2. Marinate in dressing 5 to 10 minutes. Drain.
3. Dip in beaten egg. Roll in cracker crumbs. Let dry 5 minutes.
4. Fry in hot fat. Drain, season and serve. (2 to 3 servings)

[1] Review information in Appendix G-II concerning regulations about cooking temperatures for meat and poultry.
[2] Refer to Microwave Cooking chapter for microwave recipes.

Pan-Fried Liver

beef liver	½ lb	227 g	onion, sliced	1	1
green pepper	¼	¼	oil	2 T	30 ml
cut into strips			stewed tomatoes	¾ c	180 ml

1. Cut liver into pieces.
2. Sauté liver, green pepper and onion in oil until liver is lightly brown.
3. Add tomatoes and simmer 10 minutes until just tender. Season. If desired, thicken gravy with ½ T (7.5 ml) flour. (2 to 3 servings)

SHELLFISH RECIPES

Simmered Shrimp

water	1 c	240 ml	sprig parsley	1	1
peppercorns	4	4	vinegar	1 t	5 ml
bay leaf	1	1	shrimp	½ lb	227 g
stalk celery	½	½			

1. Simmer the water with all ingredients except the shrimp for about 5 minutes.
2. Remove the shells from shrimp. With a knife, cut the shrimp just below the surface down the back. Lift out the black sand vein.
3. Add the shrimp to the stock. Cover the pan and simmer the shrimp for 5 minutes. Do not boil. Drain and chill the shrimp. Season with Cocktail Sauce or add to a curry or creole sauce.

Cocktail Sauce: Blend and chill: ¼ c (60 ml) tomato sauce, ¼ c (60 ml) chili sauce, 1 T (15 ml) grated horseradish. Yield: ½ c (120 ml).

Baked Scallops and Bacon

slices bacon	3	3	scallops	⅓ lb	151 g

1. Precook bacon until lightly brown. Remove from heat, drain, cut into 2 inch (5 cm) pieces, reserve fat.
2. Wash scallops to remove all sand. If scallops are large, cut them in half.
3. Cook scallops for 2 to 3 minutes in 2 T (30 ml) reserved bacon fat.
4. Wrap bacon pieces around scallops and place on skewers. Place skewers across baking dish.
5. Bake at 350 F (175 C) until bacon is crisp, 10 to 12 minutes (2 servings).

MEAT, POULTRY, AND FISH

FISH RECIPES[1]

Broiled Fish

| fish fillets | ½ lb | 227 g | low fat Italian dressing | ¼ c | 60 ml |

1. Turn on broiler.
2. Brush fish with dressing.
3. Place on preheated, greased broiler rack, skin side down, about 2 inches (5 cm) from the heat.
4. Broil until browned, turn with a wide spatula turner and brown the other side.
5. Baste with dressing if dry. Broil 10 to 15 minutes or until fish flakes. Season. (2 to 3 servings)

Fish Chowder

fish	⅓ lb	151 g	potatoes, diced	½ c	120 ml
water	1½ c	360 ml	milk	1 c	240 ml
fat salt pork, diced (or oil)	1 T	15 ml	evaporated milk	½ c	120 ml
chopped onions	¼ c	60 ml			

1. Clean the fish; poach fish by placing in water. Simmer, covered, until fish flakes.
2. Meanwhile, cook salt pork in a skillet until golden brown and crisp, remove and drain almost all fat; add onion and cook until soft (or sauté onion in oil).
3. Remove cooked fish from water; pour ½ c (120 ml) of cooking water over the onions, add the diced potatoes and cook until tender, about 10 minutes. Add milks.
4. Separate fish carefully into flakes and add to milk, potatoes, and onions. Simmer to reheat fish. Season and, if desired, add salt pork.

Quick Oven-Baked Fish

| fish fillets | ½ lb | 227 g | fine bread/cornflake crumbs | ½ c | 120 ml |
| milk (whole or evaporated) | ¼ c | 60 ml | margarine | 1 T | 15 ml |

1. Set oven at 500°F (260°C).
2. Dip fillets into milk. Roll in crumbs.
3. Place fish skin-side down in a greased baking pan.
4. Drizzle with melted fat.
5. Bake 10 to 12 minutes (until fish flakes). Season. (2 to 3 servings)

Spicy Baked Fish

fish fillets	½ lb	227 g	oil	2 t	10 ml
chopped onion	¼ c	60 ml	canned tomatoes	1 c	240 ml
chopped green pepper	¼ c	60 ml			

1. Set oven at 350°F (175°C).
2. Cut fish into 2 servings. Place in greased baking dish.
3. Bake until fish flakes easily, about 20 minutes. Drain liquid from fish.
4. Meanwhile, cook onion and pepper in oil until clear.
5. Cut up large pieces of tomatoes. Add tomatoes. Cook to blend flavors.
6. Pour sauce over drained fish. Bake at 350°F (175°C) for 10 minutes. Season. (2 servings)

[1] Haddock, perch, sole, flounder, turbot, or cod are suggested.

POULTRY RECIPES

Chicken Cacciatore

chopped onion	⅓ c	80 ml	garlic clove	1	1
boiling water	¼ c	60 ml	oregano	½ t	2.5 ml
canned tomatoes	1 c	240 ml	celery seed	¼ t	1.25 ml
tomato puree	¼ c	60 ml	chicken breast halves	2	2

1. Cook onion in boiling water. Do not drain.
2. Add all ingredients, except chicken. Simmer 10 minutes to blend flavors.
3. Place breast halves in heavy frying pan. Pour tomato mixture over chicken.
4. Cook, covered, over low heat until chicken is tender, about 30 to 35 minutes. (2 servings)

Oven-Fried Chicken

chicken, cut into pieces	½	½	garlic powder	dash	dash
oil or Italian dressing	2–3 T	30–45 ml			

1. Set oven at 375°F (190°C).
2. Place chicken, skin side up (or use skinless) in baking pan to which oil or dressing has been added.
3. Turn chicken to coat with oil, cook skin side down. Season.
4. Bake 40 minutes uncovered. If desired, turn chicken last 10 minutes to brown all sides. (1 to 2 servings)

Stir-Fried Chicken

boned, chicken breast, sliced	1	1	chicken broth	1 c	240 ml
peanut oil	1 T	15 ml	low-sodium soy sauce	2 T	30 ml
green pepper, sliced	¼ c	60 ml	green onion	2	2
chopped onion	¼ c	60 ml	cold water	2 t	10 ml
chopped celery	1 c	240 ml	cornstarch	2 t	10 ml
sliced mushrooms	½ c	120 ml	cooked rice	2 c	480 ml

1. Cook chicken in oil by rapidly lifting and turning the sliced pieces to expose all of the raw surfaces to the hot pan surface. Remove from skilled or push up sides of pan or wok.
2. Cook pepper and onion in skillet for a few minutes, turning constantly.
3. Add celery, mushrooms, broth and soy sauce.
4. Add chicken back to mixture, cooking 2 to 3 minutes, keeping vegetables tender crisp. Add green onion.
5. Mix cornstarch with cold water, stir into hot mixture and cook just to thicken.
6. Remove from heat and serve with hot rice. (2 servings)

MEAT, POULTRY, AND FISH

OTHER MEAT RECIPES

Breaded Veal Cutlet

veal cutlets	2	2	egg, slightly beaten	1	1
flour			fine bread crumbs	1/3 c	80 ml
oregano			oil	1–2 T	15–30 ml

1. Dip veal cutlet in seasoned flour.
2. Dip cutlet into egg, then in crumbs. Let dry 5 minutes.
3. Pan fry *slowly* in hot oil in heavy skillet (about 15 minutes each side). Serve as is, or with a tomato sauce. (2 servings)

Curry of Lamb

lamb (round)	1/3 lb	151 g	chopped green pepper	1/4 c	60 ml
margarine	2 T	30 ml	curry powder	1–2 t	5–10 ml
clove of garlic	1	1	apple, diced	1/2	1/2
finely chopped onion	1/2 c	120 ml	carrot, diced	1/2	1/2
chopped celery	1/4 c	60 ml	water	1/4 c	60 ml

1. Cut meat into 1- to 2-inch (2.5- to 5-cm) pieces, and brown lightly in fat.
2. Push meat to one side of frying pan. Add garlic, onion, celery and green pepper to pan, and cook slowly for 2 to 3 minutes. Remove garlic.
3. Add remaining ingredients, and mix with meat.
4. Cover and simmer until meat is tender, about 1 hour. NOTE: Cooked lamb, beef, chicken, or fish, or uncooked beef shoulder or round may be used. (2 servings)

Broiled Steak
(Sirloin, T-bone, Rib, or Porterhouse Steak)

1. Grease broiler rack and preheat broiler. (Follow oven directions on use of broiler.)
2. Slash the outer fat of the meat in several places to prevent curling.
3. After allowing meat to stand at room temperature for a few minutes, place meat on broiler rack.
4. Broil 3 to 4 inches (7.5 to 10 cm) from heat. Rate of heat is regulated by distance meat is placed from heating unit.
5. Turn steak when upper side is brown. Season to taste.

Broiled Flank—London Broil

1. Grease broiler rack and preheat broiler.
2. Allow meat to set out a few minutes at room temperature. Score meat by cutting diagonally across the long muscle fibers.
3. Place steak on rack. Broil 2 to 3 inches (5 to 7.5 cm) from heat about 5 minutes on each side. Cook to rare. Carve across grain.

Picture Courtesy of: National Cattlemen's Beef Association

Pan-Broiled Steak

1. Slash the outer fat edge in several places to prevent curling.
2. Place meat in heavy frying pan.
3. Cook, uncovered, slowly. Do not add fat or water.
4. Pour fat from pan as it accumulates.
5. Turn meat to brown other side.
6. Cook to desired doneness. Season to taste.

Pan-Broiled Pork Chops

1. Place 2 pork chops in heavy frying pan.
2. Over moderate heat, brown each side of chop.
3. Reduce heat. Cook chops slowly, uncovered.
4. Pour off excess grease as it accumulates.
5. Cook until chops are well done. Season. (2 servings)

Braised Pork Chops

pork chops	2	2	chili powder	¼ t	1.25 ml
oil	1 T	15 ml	chopped onion	1 T	15 ml
tomato sauce	½ c	120 ml	vinegar	1½ T	22.5 ml
Worcestershire sauce	1 t	5 ml			

1. Brown chops in oil. Pour off excess fat.
2. Combine remaining ingredients. Add to chops.
3. Cook, covered, 25 to 30 minutes, or until chops are done. (2 servings)

SUMMARY QUESTIONS—MEAT, POULTRY, AND FISH

1. Make a summary statement concerning the relationship of the inherent tenderness of a cut of meat and its price. Discuss whether a similar generalization can be applied to the comparison of a cut of meat and its nutritive value.

2. List four "dry heat" methods of cooking.

3. List four "moist heat" methods of cooking.

Picture Courtesy of: National Cattlemen's Beef Association

4. List cuts of meat that are often breaded. What method of cooking is used for breaded products?

5. Predict the outcome as to relative juiciness and tenderness, if a pressure cooker was used to stew a bottom round. Explain.

6. Select an appropriate method of cooking for each of the following products. Explain.

	Method	Explanation
Heart		
Liver		
Cod fillets		
Shrimp		
Fowl (mature hen)		

7. Based on laboratory observations and/or readings, discuss internal cooking temperatures of meat in relation to palatability characteristics of a meat product.

Degree of Doneness	Internal Temperature	Palatability
Rare		
Medium		
Well done		

8. Complete the following nutritive value chart.

Food Item 3 oz (90 g)	Energy (kcal)	Protein (g)	Fat (g)	Cholesterol (mg)	Iron (g)
Beef liver					
Ground beef					
Chicken breast					
Haddock					
Pork chop					
Veal					

9. In summary, identify the major nutrient contributions of meat, poultry, and fish.

10. What nutrients in meats are adversely affected by long cooking and high cooking temperatures?

11. Turkey is ground for substitution in ground meat recipes, or it may be processed into lunch meat and hot dogs. What are the advantages of using turkey meat?

12. When a large roast is removed from the oven, what occurs to the internal temperature? Explain. Based on this fact, how can a roast be cooked to only the medium-done stage?

13. Recipes for roasting turkey over 15 pounds (6.8 kg) include directions to bake stuffing separately. Why is separate roasting necessary?

14. Complete the following chart: (see Appendix G-II)

Meat	End Cooking Temperature	Microorganism Needing Control
Beef		
Pork		
Chicken		
Ground beef patty		

15. Why does ground beef have a shorter shelf life than the roast from which it came? What other flesh products are considered highly perishable?

16. List several precautions that should be taken in the use of cutting boards and knives that have been used to cut items such as chicken, beef, pork, or fish.

17. Apply principles of time and temperature interaction to the preservation of the sanitary quality of meat.

18. Explain why "Quick Oven Baked Fish" is tender and juicy, yet the 500°F (260°C) oven temperature appears to contradict the principle of using low temperatures for protein foods.

19. At current prices, which of the following would be the best protein buy per serving: beef liver? hot dogs? hamburger? haddock?

20. Relate the following graph to the cooking of an inherently tough and inherently tender cut of meat.

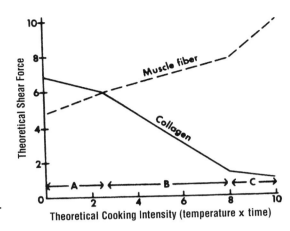

THEORETICAL COOKING INTENSITY. EFFECT OF COOKING ON TENDERNESS OF MEAT.

Source: Wang HH, et al. 1954. Food Research ® Institute of Food Technologists. Reprinted with permission.

a. For which cut would an increasingly "tough, dry" product be expected over time? Explain.

b. For which cut would greater cooking intensity be advantageous? Explain.

c. Select one recipe from the meat unit and explain how cooking intensity (time and temperature) was used to achieve a palatable product.

21. Identify "Safe Handling Instructions" that appear on meat packages.

E. Plant Proteins

OBJECTIVES

To demonstrate how to rehydrate and cook legumes and to understand principles involved in achieving a palatable product

To identify the nutritive value of grains, legumes and seeds as meat alternatives

To identify various combinations of grains, legumes and seeds as well as combinations with milk that provide a complete amino acid pattern

To become familiar with a variety of meat alternatives by incorporating them into palatable, nutritious recipes

To appraise the nutritive, sanitary and economic dimensions of plant proteins

REFERENCES

Appendices N, O

ASSIGNED READINGS

128 Food Principles

TERMS

Lentils
Legumes
Pectin
Complete protein
Incomplete protein
Partially complete protein
PER
Essential amino acid
Mutual supplementation
Lysine
Tryptophan
Sulfur containing amino acids
Lacto-ovo vegetarian
Vegetarian, vegan

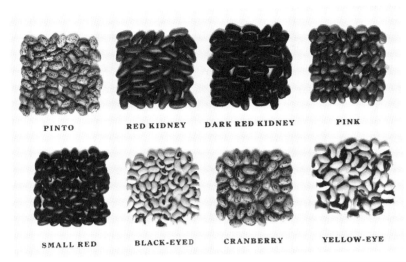

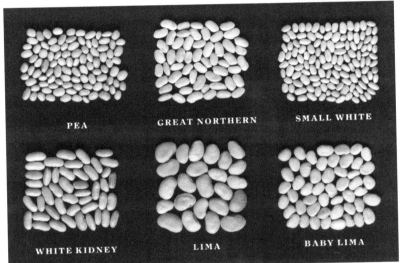

PRINCIPAL TYPES OF BEANS. (TOP) DRY COLORED BEANS; (BOTTOM) DRY WHITE BEANS
Courtesy: USDA

EXERCISE 1: PRETREATMENT AND COOKING METHODS FOR LEGUMES/LENTILS

A. PRETREATMENT

PROCEDURE

1. Add 1½ c (360 ml) water to ½ c (120 ml) assigned beans. Bring to a boil and boil 2 to 3 minutes. Turn off heat and soak, covered for 1 hour.

    AND/OR

2. Add 1½ c (360 ml) water to assigned beans, cover, refrigerate overnight.

B. Cooking Methods

Legumes: Place pretreated beans in uncovered saucepan, adding water to cover if necessary, and boil. Lower heat and simmer until beans are tender.

Lentils/split peas: Place ½ c (120 ml) unsoaked lentils or split peas in saucepan and cover with water. Cook until tender.

Procedure

Record time and cooked yield. If using in Exercise 2, place in covered containers and refrigerate or freeze.

Evaluation of Cooked Legumes, Lentils

	Cooking Time	Cooked Yield	Palatability
Black-eyed peas			
Garbanzo beans (chick peas)			
Great Northern beans			
Lentils			
Lima beans			
Navy beans			
Red kidney beans			
Soybeans			
Split peas			

VEGETABLE PROTEIN CASSEROLE

Source: Division of Nutritional Sciences, New York State College of Human Ecology at Cornell

EXERCISE 2: COMBINING PLANT PROTEINS

PROCEDURE

1. Follow directions for assigned product. Plan to serve in 1 hour.
2. Display product and evaluate all finished products for palatability, nutritive value, and general acceptability.

EVALUATION OF PLANT PROTEIN RECIPES

Recipe	Plant Protein	Palatability			Comments
		Appearance	Texture	Flavor	

PLANT PROTEIN RECIPES[1]

Frijoles (Beans)

cooked red or pinto beans	2½ c	600 ml	cooked tomatoes	1 c	240 ml
chili powder	½ t	2.5 ml	chopped celery	⅓ c	80 ml
cayenne pepper	⅛ t	.63 ml	cooked rice	2 c	480 ml
chopped onion	⅓ c	80 ml			

1. Combine all ingredients except rice and simmer, covered about 30 minutes. Stir occasionally. Adjust seasoning.
2. Serve over rice. (2 to 3 servings)

Beans and Rice Casserole

cooked beans (garbanzo, red beans, etc.)	2¼ c	660 ml	green pepper, chopped	¼ c	60 ml
			tomato sauce	¾ c	180 ml
oil	½ T	7.5 ml	basil	½ t	2.5 ml
finely chopped onion	¼ c	60 ml	oregano	½ t	2.5 ml
finely chopped carrots	2	2	cooked rice	1 c	240 ml
chopped celery	½ c	120 ml	grated cheese	¼ c	60 ml

1. In a large skillet, sauté onion, carrots, celery, and pepper in oil until softened.
2. Add beans, tomato sauce and seasonings; simmer.
3. Combine rice and bean mixture, or spoon bean mixture over rice. Sprinkle with cheese. (2 to 3 servings)

Bean Salad

cooked wax beans	⅓ c	80 ml	vinegar	1 T	15 ml
cooked green beans	⅓ c	80 ml	oil	2 T	30 ml
cooked kidney beans	⅓ c	80 ml	sugar	½ t	2.5 ml
finely chopped onion	1 T	15 ml	lettuce leaves	2–3	2–3
finely chopped celery	2 T	30 ml			

1. Mix beans and vegetables.
2. Beat vinegar, oil and sugar. Add to vegetables, mixing gently.
3. Refrigerate for at least 1 hour. Spoon mixture onto lettuce leaves. Season. (2 to 3 servings)

[1] Cooked (canned) beans are used in these recipes to conserve preparation time. When using canned beans, drain and wash to reduce salt.

PLANT PROTEINS 133

Many Bean Soup

chopped celery	½ c	120 ml	liquid from beans or water	1 c	240 ml
chopped onion	⅓ c	80 ml	cooked beans	2 c	480 ml
oil	1 T	15 ml	(navy, kidney, lima, etc.)		
sliced carrots	1 c	240 ml	cooked tomatoes, mashed	½ c	120 ml
medium potato, diced	1	1	dill weed	½ t	2.5 ml

1. In a medium saucepan, sauté celery and onion in oil until soft.
2. Add carrots, potatoes and liquid. Boil until vegetables are just tender.
3. Add beans, tomatoes and dill weed. Simmer gently until mixture is heated through. Season. Serve with a grain product. (4 servings)

Cornmeal-Bean Bread

oil	1 T	15 ml	stock from beans	1 c	240 ml
chopped onion	¼ c	60 ml	(or 1 c (240 ml) beef		
cornmeal	1 c	240 ml	bouillon)	1	1
baking powder	2 t	10 ml	egg, beaten	¼–⅓ c	60–80 ml
chili powder	½ T	7.5 ml	grated cheese	¼ c	60 ml
cooked kidney beans, chopped	1½ c	360 ml	sliced black olives		

1. Set oven at 350°F (175°C).
2. Sauté onion in oil in medium skillet. Remove onion and reserve.
3. Mix cornmeal, baking powder and seasonings in a bowl.
4. Combine onion, kidney beans, stock and egg. Add to dry ingredients mixing just to moisten.
5. Pour mixture into skillet. Sprinkle with cheese and olives. Season.
6. Bake 15 minutes or until bread tests done. (4 servings)

Eggplant Casserole

cooked tomatoes, drained	1 c	240 ml	sesame seed	¼ c	60 ml
oregano	½ t	2.5 ml	large eggplant, peeled, sliced	½	½
thyme	¼ t	1.25 ml	oil	2 t	10 ml
finely chopped onion	2 T	30 ml	mozzarella cheese, sliced	¼ lb	114 g
chopped green pepper	2 T	30 ml	cooked rice	1½ c	360 ml
grated cheese (parmesan or cheddar)	¼ c	60 ml			

1. Set oven at 350°F (175°C).
2. Combine tomatoes, seasonings, onion and green pepper. Cover and simmer 10 to 15 minutes. Add grated cheese and sesame seed.
3. Meanwhile, using a large frying pan, sauté eggplant slices in oil until lightly browned. Drain.
4. Place a layer of eggplant in a 2-quart (2-L) greased baking dish. Cover with half the tomato sauce and half the mozzarella cheese. Repeat layers.
5. Bake about 30 minutes until cheese browns. Serve with rice. Season. (3 servings)

Enchiladas-Bean and Cheese

vegetable oil	3 T	45 ml	water	3 c	720 ml
flour	3 T	45 ml	refried beans	1½ c	360 ml
chili powder	3 T	45 ml	corn tortillas	12	
tomato bouillon with chicken flavor	1 T	15 ml	jack cheese	½ c	120 ml

1. Brown flour. Add chili powder and oil.
2. Add water and bouillon. Stir. Bring to a boil until sauce thickens, stirring well.
3. Dip tortillas into sauce. Fill each tortilla with 1 T (15 ml) heated refried beans and roll.
4. Arrange in a casserole dish and pour remaining sauce on top. Sprinkle with cheese.
5. Place in 350°F (175°C) oven 10 to 15 minutes to melt cheese.

Hoppin' John

bacon, slice	1	1	cooked black beans	1 c	240 ml
finely chopped onion	¼ c	60 ml	cooked rice	1 c	240 ml
garlic clove, minced (opt)	1	1	water or chicken stock	⅓ c	80 ml

1. Fry bacon, onion and garlic in large saucepan. Remove bacon when crisp, drain and crumble. Reserve.
2. Add beans, rice and water to fat.
3. Bring to boil, lower heat and simmer 10 minutes. Season and add bacon. (2 to 3 servings)

Hummus (Garbanzo–Tahini Spread) with Pita Bread

onion, minced	1 lg.		lemon juice, fresh	½ c	120 ml
garlic, minced clove	1		soy sauce, reduced-sodium	1 T	15 ml
oil, vegetable	1 T	15 ml	sesame paste (Tahini)	¼ c	60 ml
garbanzo beans	2 c	480 ml	pitas	4	

1. Sauté onion and garlic until onion is soft.
2. Using a blender, puree all ingredients. Serve with pita bread or as a vegetable dip. Yield: 2 cups.

Lentil Burgers

dry lentils	¾ c	180 ml	eggs, slightly beaten	2	2
water	1½ c	360 ml	oregano	¼ t	1.25 ml
finely chopped onion	⅓ c	80 ml	oil	2 T	30 ml
grated carrots	½ c	120 ml	processed cheese (opt)	2 sl	2 sl
dry bread crumbs (whole wheat)	1½ c	360 ml			

1. Add water to the lentils; bring to boil. Cover and simmer 15 minutes.
2. Add onion and carrots; cook about 15 minutes or until lentils are tender.
3. Cool slightly. Add crumbs, eggs and oregano. Mix well.
4. Heat oil in large skillet. Drop lentil mixture ½ cup (120 ml) at a time into hot oil. Flatten to make patties.
5. Cook patties until firm, about 7 minutes on each side. If desired, top each pattie with cheese and heat to melt cheese. (4 servings)

Soybean–Corn–Tomato Casserole

cooked soybeans	2 c	480 ml	garlic powder	¼ t	1.25 ml
whole kernel corn, drained	1 c	240 ml	oregano	¼ t	1.25 ml
cooked tomatoes	1 c	240 ml	basil leaves	½ t	2.5 ml
flour	1 t	5 ml	cheese, shredded	1 oz	28 g

1. Set oven at 375°F (190°C).
2. Arrange beans and corn in alternate layers in a 1-quart (1-L) greased baking dish.
3. Mash tomatoes with a fork; reserve 2 T (30 ml) tomato juice.
4. Mix flour and seasonings in small saucepan. Combine reserved tomato juice and flour, add to tomatoes.
5. Heat, stirring until mixture boils.
6. Pour hot sauce over vegetables and bake about 20 minutes until heated through. The last 5 minutes, sprinkle with cheese. Season. (4 servings)

Stuffed Peppers

oil	1 T	15 ml	cooked beans, mashed	1½ c	360 ml
onion, chopped	¼ c	60 ml	(e.g., kidney, pea, garbanzo)		
celery, chopped	¼ c	60 ml	basil	½ t	2.5 ml
cooked tomatoes	1 c	240 ml	cheddar cheese, grated	⅓ c	80 ml
			green peppers, seeded,	3	3

1. Set oven at 400°F (205°C).
2. Sauté onion and celery in oil until onion is lightly browned.
3. Add tomatoes, beans and seasonings. Remove from heat and add cheese.
4. Fill the pepper halves with mixture.
5. Place peppers in an oblong baking pan with about 1 inch (2.54 ml) hot water in bottom of the pan. Cover.
6. Bake 15 minutes; uncover and bake 10 to 15 minutes longer until peppers are just tender. Keep water in the pan. Season. (3 servings)

Tamale Pie

Filling:

margarine	½ T	7.5 ml	meat stock	¼ c	60 ml
diced onion	¼ c	60 ml	chili powder	¾ t	4 ml
cooked kidney beans,	1 c	240 ml	chopped olives (optional)	2 T	30 ml
tomato soup	⅜ c	90 ml	grated cheddar cheese	2 T	30 ml

Cornbread Topping:

flour	2 T	30 ml	buttermilk	¼ c	60 ml
baking soda	⅛ t	.63 ml	egg, beaten	½	½
cornmeal	3 T	45 ml	melted margarine	1 T	15 ml

1. Set oven at 425°F (220°C).
2. Brown the onion in the fat. Add all ingredients except the cheese and simmer for about 5 minutes. Pour into two 6-oz (180-ml) greased custard dishes. Sprinkle with cheese.
3. Sift the flour, soda and salt together. Mix the cornmeal with the dry ingredients.
4. Combine the buttermilk, beaten egg and melted fat and add to dry ingredients. Mix just to moisten. Spread batter over bean mixture.
5. Bake 20 minutes or until cornbread is golden brown. Season. (2 servings)

TOFU RECIPES[1]

Tofu Burgers

tofu	6 oz	170 g	parmesan cheese, grated	2 t	10 ml
egg	1	1	pepper	¼ t	1.25 ml
bread crumbs	½ c	120 ml	oregano	¼ t	1.25 ml
onion, minced	2 T	30 ml	cayenne	dash	dash
garlic, minced	1 t	5 ml	oil, vegetable	1 T	15 ml

1. Combine all ingredients in a bowl and stir until well mixed.
2. Form the tofu mixture into 4 patties and fry to brown both sides.
3. Heat oil in 10- to 12-inch skillet.
4. Place on baking sheet in 350°F (175°C) oven for 10 to 15 minutes. If desired, serve with lettuce and tomato in a bun. (4 servings)

Stir-Fried Tofu with Spinach

raw rice	1 c	240 ml	chopped spinach	5 oz	142 g
tofu	½ lb	227 g	low-sodium soy sauce	½–1 T	7–15 ml
peanut oil	2 t	10 ml			

1. Cook rice. Maintain temperature at or above 140°F (60°C).
2. In a large skillet sauté tofu cubes in oil about 5 minutes. Stir gently. Push cubes to center and spread spinach around edge.
3. Sprinkle with soy sauce and cover. Steam mixture until spinach has just wilted.
4. Season. Serve mixture over hot rice. (2 servings)

Pineapple Banana Shake

soft tofu	4–6 oz	114–170 g	orange juice	½ c	120 ml
crushed pineapple	1 c	240 ml	banana	½	½

Place all ingredients in blender and process until smooth. If too thick, add more fruit juice. Keep refrigerated. Serve chilled. (2 c [480 ml])

---

[1] Soybean curd (Tofu) is coagulated soy protein. Fresh curd, which is perishable, should be cut and the liquid drained off before adding to any recipe.

SUMMARY QUESTIONS—PLANT PROTEINS

1. In cooking dried beans:
   a. What steps are important in obtaining a tender product?

   b. What common ingredients, if added too early in the cooking process, will cause the beans to harden?

   c. In cooking beans, how are the pectin, protein and starch of the beans changed?

2. Regarding the use of dried beans:
   a. One cup of dried beans is equal to how many cups of cooked beans?

   b. Contrast the cost of dried beans to canned. When might canned products be advantageous?

3. One cup of cooked, dried beans has approximately how many grams of protein? What other major nutrients do legumes contribute?

4. Complete the following chart:

	Amino Acids[a]	
	High	Low
Grains		
Legumes		
Soybeans		
Nuts		
Seeds		
Lentils		

[a]Consult Appendix O.

5. List several commonly used food combinations that illustrate the principle of mutual supplementation of proteins.

6. Explore several cookbooks of other regions in the United States as well as other countries.
   a. List vegetable protein dishes and characteristic meals from each region/country.

   b. Discuss how principles of mutual supplementation of proteins have been applied in these dishes and meals.

   c. Evaluate the potential protein quality of these vegetable protein dishes.

7. Discuss other nutritional dimensions (e.g., minerals, vitamins) that should be considered when substituting plant protein for animal protein in the diet.

8. Why are legumes considered the foundation of a *strict* vegetarian diet?

9. A friend is switching to a lacto-ovo vegetarian diet after 20 years on a "typical American diet." Summarize key points you would suggest about the nutritional quality of the new diet. How would advice differ if the friend was changing to a strict vegetarian regimen?

F. Eggs

OBJECTIVES

To identify characteristics indicative of egg quality and relate these to use of eggs in food preparation
To observe the time-temperature relationships that occur during the coagulation of egg proteins
To know and apply temperature standards for safe handling of cooked egg products.
To describe the effect of manipulation, especially stirring and rate of heating on the coagulation temperature of egg mixtures
To describe the effect of ingredients and their proportions on the coagulation of egg mixtures
To demonstrate preparation of an egg white foam
To delineate factors that affect both foam volume and stability
To relate egg characteristics to uses of eggs in food preparation
To apply principles of the combination of starch and egg cookery in food preparation
To appraise the nutritive, sanitary and economic dimensions of eggs and egg substitutes

REFERENCES

ASSIGNED READINGS

Appendices G-I, G-II; N

TERMS

Coagulation	Curdling	Stiff peak	Bake
Coagulation temperature	Syneresis	Dry peak	Poach
Intrabonding	Weeping	Whip	Fry
Interbonding	Foam, foamy	Beat	Egg substitutes
Sol	Soft peak	Fold	*Salmonella*
Gel			

 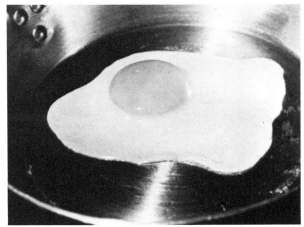

(LEFT) FRESH EGG WITH UPSTANDING YOLK AND FIRM WHITE CONTRASTED TO OLDER EGG (RIGHT)

Courtesy: USDA

EXERCISE 1: EGG QUALITY

PROCEDURE

1. Place eggs of various freshness (age) in a bowl of water.
2. Note which eggs float, and which sink.
3. Carefully open each egg and place each in a saucer. Observe characteristics of the white and yolk.
4. Record observations and summarize conclusions.

Whole Egg in Water	Description of White	Description of Yolk

Conclusions:

1. Why did some eggs float? Explain.

2. What other methods are used for judging the quality of eggs?

EXERCISE 2: COAGULATION OF EGG PROTEIN IN BAKED AND STIRRED CUSTARD

BASIC EGG CUSTARD

milk	2 c	480 ml	sugar		2 T	30 ml
eggs, large	2	2	vanilla		½ t	2.5 ml

PROCEDURE

1. Calibrate thermometer and set oven at 400°F (205°C). Label small paper cups for stirred custard samples.
2. Scald milk in top of double boiler over hot water.
3. Place egg and sugar into a medium size mixing bowl and mix slightly. Pour scalded milk *slowly* into the mixture, *stirring constantly*. Add vanilla. Use the mixture for BOTH the Baked and Stirred Custard preparation as directed.

BAKED CUSTARD[1]

1. Fill 2 custard cups ¾ full of the BASIC EGG CUSTARD recipe. Reserve the remainder of the mixture for the Stirred Custard.
2. Place one cup in a pan of hot water, with water level even with the custard.
3. Place the second cup in a pan or on a cookie sheet with no water.
4. Bake for 25 minutes, or until the sample cooked in water is fully cooked. (Fully cooked is determined by inserting a metal knife into the custard, halfway between the edge and center. When the knife comes out clean, the custard is done.)
5. Label and cool the baked samples. Record observations of appearance, texture, and consistency, using palatability terms provided.

[1] Refer to Microwave Cooking chapter for microwave recipes.

TESTING EGG CUSTARD

Courtesy: American Egg Board

BAKED CUSTARD

	Appearance	Texture	Consistency
Baked in water bath			
Baked without water			
Baked in microwave			

Conclusions:

STIRRED CUSTARD:

1. Pour the remainder of the BASIC EGG CUSTARD recipe mixture (reserved from the **Baked** Egg Custard preparation) back into the top of the double boiler.
2. Place cold water in the bottom of the double boiler at a level that does not touch the top pan.
3. Begin to cook the mixture, stirring as soon as it is placed over the heat.
4. After the water on the bottom starts to boil, turn the heat to low to maintain a simmer. Do not boil.
5. Hold the calibrated thermometer in the center of the mixture contents, *resting it on the bottom of the top pan*. Do not remove the thermometer while the mixture is cooking.
6. Stir the mixture continuously, while *quickly* removing samples (placing samples in paper cups) with a metal spoon at designated temperatures.
7. While removing samples, note at what temperatures the metal spoon is slightly coated and at what temperature the coating becomes heavy and velvety. Record observations at each temperature.
8. After the custard has reached maximum thickness, continue to cook it until it curdles. Record the curdling temperature.
9. Label and cool all samples. Record observations and summarize the effect of increasing temperature on egg protein, appearance, texture, and consistency, using appropriate palatability terms.

PALATABILITY TERMS		
APPEARANCE	**TEXTURE**	**CONSISTENCY**
Glossy	Smooth	Thin
Dull	Velvety	Watery
Shiny	Lumpy	Slightly thickened
	Curdled	Thick
	Porous	Gel-like
		Firm

STIRRED CUSTARD

Sample Temperature	Appearance	Texture	Consistency
177°F (81°C)			
181.4°F (83°C)			
183°F (84°C)			
185°F (85°C)			
187°F (86°C)			
188.6°F (87°C)			
190.4°F (88°C)			
Curdled			

Conclusions:

1. Protein is the component responsible for the functional properties of eggs in food preparation.
   a. Describe how protein is dispersed in a raw egg (sol/gel).

   b. Describe how protein is dispersed in a cooked egg (sol/gel).

   c. Explain how heat, including high heat affects egg protein.

   d. At what temperature do egg yolks and egg whites coagulate?

2. What changes in egg protein structure take place while the egg mixture is cooking:
   a. If prepared as a stirred custard?

   b. If prepared as a baked custard?

3. Describe what occurs in terms of structure, when the temperature of the custard is raised above the coagulation point of the egg protein.

4. How does the speed of cooking a stirred custard affect coagulation temperature? Explain any adverse affect speed of cooking has on the product.

5. Does a fully cooked baked custard become appreciably thicker upon cooling? Explain.

6. What is the purpose of using a water bath when baking custard? For what other products would a water bath be beneficial?

EXERCISE 3: EGG WHITE FOAMS

PROCEDURE
1. Separate 2 eggs, placing whites and yolks into separate bowls.
2. Beat egg whites with rotary or electric beater to designate stage of foam. Record observations on foam development as beating continues.
3. Beat yolks until *thick and lemon-colored*, then fold yolks into beaten whites.
4. Summarize observations; contrast effect of beating egg whites and yolks.

Stage of Foam	Volume	Description	Comments
Coarse foam			
Foamy			
Soft peak			
Stiff peak			
Dry foam			

Conclusions:

1. Describe the structural formation of the protein in an egg white foam.

2. What properties or characteristics of egg whites make them useful as leavening agents?

3. Is a foam beaten to the dry stage as effective a leavening agent as a stiff peak foam? Explain.

4. Describe the process of *folding* beaten yolks into beaten whites.

5. Contrast volume obtained by beating egg yolks until thick, to that of egg white beaten to stiff peaks. Account for differences.

Egg Component	Volume	Explanation
Beaten yolks		
Whites beaten to stiff peaks		

EXERCISE 4: EFFECT OF ADDED SUBSTANCES ON EGG WHITE FOAM

PROCEDURE

1. Measure the volume of 1 egg white, place it in a 1 quart (or liter) bowl.
2. Add one of the assigned ingredients to the white, as directed.
3. Beat each mixture for 2 minutes, or until the foam reaches a stiff peak stage.
4. Carefully measure the final volume and display the foam.
5. Record observation on foam volume, stability and general appearance.
6. Hold samples for 10 minutes and re-evaluate the foam characteristics.

Ingredient	Initial Volume	Final Volume	Stability	Comments
¼ t (1.25 ml) cream of tartar, added initially				
¼ t (1.25 ml) cream of tartar, added at foamy stage				
2 T (30 ml) sugar, added initially				
2 T (30 ml) sugar, added at soft peak stage				
1 T (15 ml) water, added initially				
⅛ t (.63 ml) oil, added initially				

Conclusions:

EXERCISE 5: EFFECT OF COOKING INTENSITY ON EGG PROTEIN

Procedure

Cook refrigerated egg(s) according to assigned procedure listed below and using assigned *cooking intensity* (time × temperature) as directed. Check final temperature of product and record palatability evaluations.

Methods of Cooking Eggs[1]

Eggs Cooked in Shell (Soft or Hard Cooked): Place whole egg in small saucepan, with water to cover.

Eggs Cooked in Water (Poached): Fill shallow pan with water, twice the depth of egg. Bring water to specified temperature. Remove egg from shell and place in custard cup. Swirl water with a spoon and carefully drop egg into vortex.

Eggs Cooked in Fat (Fried): Place 1 t (5 ml) fat in a small frying pan, melt over low heat. Remove egg from shell and place in pan.

Egg Mixture (Scrambled): Mix 2 eggs and 2 T (30 ml) milk. If scrambling in frying pan, add 1 t (5 ml) margarine and melt or use nonstick spray. Add egg mixture. If using double boiler, place egg mixture in top over simmering water in bottom pan. Gently stir mixture until it is firm and moist.

Baked Eggs (Shirred): Break egg into lightly greased custard cup. Season as desired and add ½ t (2.5 ml) margarine to egg. Bake uncovered at 350°F (175°C).

Ensuring the Safety of Eggs

1. Keep eggs refrigerated until ready to use, except for egg whites used for foams in baked products.
2. Check final temperature of cooked product.
3. DO NOT taste any egg product that does not reach a final cooking temperature of 140°F (60°C) held for 3½ minutes, or 160°F (71°C).

---

[1] Refer to Microwave Cooking chapter for microwave recipes.

Evaluation[a] of Eggs Cooked in Various Ways

Procedure	Cooking Time[a] (min)	Final Temperature	Appearance		Texture/ Tenderness
			White	Yolk	
Eggs cooked in shell 　Simmer	15				
Simmer	25				
Boil	15				
Boil	20				
Heat to boiling. Turn off heat. Stand.	15				
Heat to boiling. Turn off heat. Stand.	25				
Eggs cooked in water (poach) 　Simmer	7				
Simmer: water + 1 t (5 ml) vinegar	7				
Boil	7				
Eggs cooked in fat (fried) 　LOW–MED heat, uncovered	5				
HIGH heat, uncovered	4				
LOW heat, covered + 1 T (15 ml) water	5				
Egg mixtures (scrambled) 　Frying pan: LOW heat	3–5				
Frying pan: HIGH heat	4				
Double boiler: until firm	6–8				
Baked eggs (shirred) 　Bake	10–15				
Bake	25				

[a]Do not taste eggs that have not reached 140°F (60°C), held for 3½ minutes, or 160°F (71°C). For undercooked eggs, observe texture.

EXERCISE 6: CHARACTERISTICS OF COOKED MODIFIED EGG MIXTURES

PROCEDURE

1. Using standard procedures, pan fry and scramble the following eggs or egg mixtures that reduce or eliminate egg yolks.
2. Record observations of texture, flavor, appearance and nutritive value.

Pan fry:

    a. Two whole eggs

    b. Two egg whites only

    c. ½ c (120 ml) cholesterol-free egg product.

Scramble:

    a. Two whole eggs, 1 T (15 ml) skim milk

    b. Three egg whites + 1 egg yolk, 1 T (15 ml) skim milk

    c. ½ c (120 ml) cholesterol-free egg product

Egg	Texture	Flavor	Appearance	Cholesterol (mg)	Fat (g)	Protein (g)
Pan-fried egg						
a. Whole						
b. Egg whites						
c. Cholesterol-free						
Scrambled						
a. Whole + milk						
b. Whites + yolk + milk						
c. Cholesterol-free						

NOTE: observe temperature precautions for tasting cooked eggs.

EXERCISE 7: COMBINING STARCH AND EGGS AS THICKENERS IN ONE PRODUCT—SOUFFLÉ

Procedure

1. Prepare a soufflé that uses both starch and egg protein for thickening and structure, and egg white for leavening.
2. Evaluate the palatability of soufflé based on the following criteria.

	Palatability
Texture	
Consistency	
Volume	
Flavor	

3. Analyze the starch and protein component of the product structure (sol, gel).

	Uncooked product	Hot, cooked product
Starch		
Protein Egg white		
Egg yolk		

1. Explain how the principles of *starch* gelatinization were applied in preparing the starch/egg product.

2. Explain how the principles of *egg* protein coagulation were applied.

CHEESE SOUFFLE

Courtesy: American Egg Board

SOUFFLÉ RECIPES

Cheese Soufflé

margarine	2 T	30 ml	cheddar cheese, grated	½ c	120 ml
flour	1½ T	22.5 ml	eggs, separated	2	2
milk	½ c	120 ml			

1. Set oven at 325°F (163°C).
2. Prepare a thickened sauce of the fat, flour and milk.
3. Add the cheese and stir the mixture over low heat until cheese melts.
4. In a small bowl, beat egg yolk slightly with a fork.
5. Slowly add a little of the hot mixture to yolks, stirring to blend. Then add the warmed egg mixture to the cheese sauce. Mix thoroughly. Set aside to cool slightly.
6. Beat the egg whites until stiff.
7. *Gradually* fold the cheese–yolk mixture into the beaten whites.
8. Pour the mixture into two 10-oz (300-ml) ungreased baking dishes. To make a "high hat" on a soufflé, use a knife point and trace a circular groove on the top of mixture, about 1 inch (2.54 cm) from edge.
9. Bake 20 minutes or until a knife inserted in the center comes out clean. If a water bath is used, set oven at 375°F (190°C). (2 servings)

Chili Rellenos

green chilis, seeded deveined and chopped	4 oz	114 g	eggs, separated	2	2
			evaporated milk	½ c	120 ml
jack cheese, grated	⅓ lb	151 g	flour	2 T	30 ml

1. Set oven at 325°F (165°C).
2. Place a layer of chilis in three 10-oz (300-ml) greased baking dishes.
3. Cover with a layer of cheese. Repeat.
4. Beat egg whites until stiff.
5. Beat evaporated milk, flour, and egg yolks until well blended.
6. Gently fold egg whites into yolk mixture (mixture is thin).
7. Pour egg–milk mixture over chilis.
8. Bake 25 to 30 minutes or until custard has set. (3 servings)

Chocolate Soufflé

flour	½ T	7.5 ml	margarine (optional)	1 t	5 ml
sugar	3 T	45 ml	vanilla	1 t	5 ml
milk	½ c	120 ml	eggs, separated	2	2
baking chocolate, grated	½ oz	14 g			

1. Set oven at 325°F (165°C).
2. Mix flour and sugar in saucepan. Add milk and chocolate, stirring.
3. Heat, stirring constantly until thickened. Stir in margarine and vanilla. Set aside.
4. Beat yolks until thick and lemon colored. Slowly add the thickened milk to yolks, stirring constantly.
5. Beat egg whites until stiff.
6. Fold chocolate mixture into whites. Pour into three 6 oz (180 ml) ungreased baking dishes.
7. Bake 25 to 30 minutes, or until custard has set. (3 servings)

Spoon Bread

milk	1½ c	360 ml	sugar	1 T	15 ml
cornmeal, white	⅓ c	80 ml	baking powder	½ t	2.5 ml
margarine	1 T	15 ml	eggs, separated	2	2

1. Set oven at 325°F (165°C).
2. Scald milk in top of double boiler.
3. Stir in cornmeal gradually and cook over boiling water until thickened, stirring occasionally. If very thick, add ¼ c (60 ml) milk.
4. Add fat, sugar and baking powder; mix well.
5. Beat egg yolks until thick and lemon colored. Add hot cornmeal mixture slowly to beaten egg yolks, stirring to blend.
6. Beat egg whites until stiff.
7. Gently fold cornmeal mixture into beaten whites.
8. Spoon into six 4-oz (120-ml) ungreased baking dishes.
9. Bake 30 minutes, or until custard has set. (6 servings)

Tuna Soufflé

margarine	1½ T	22.5 ml	tuna, drained and flaked[1]	3 oz	85 g
onion, minced	1 t	5 ml	paprika	dash	dash
flour	1½ T	22.5 ml	eggs, separated	2	2
milk	½ c	120 ml			

1. Set oven at 325°F (165°C).
2. Melt fat and sauté onion. Blend in flour.
3. Add milk gradually and cook, stirring constantly until mixture thickens. Remove from heat and stir in tuna and paprika.
4. Beat egg yolks until thick and lemon colored. Slowly add hot tuna mixture to yolks, stirring constantly.
5. Beat egg whites until stiff.
6. Gently fold tuna-yolk mixture into beaten whites. Spoon into two 10-oz (300-ml) or four 4-oz (120-ml) ungreased baking dishes.
7. Bake 30 minutes, or until custard has set. (3 servings)

[1]Canned salmon, chopped cooked shrimp, chicken, or ham may be used. The addition of ¼ c (60 ml) grated cheddar cheese enhances flavor.

Vegetable Soufflé

margarine	2 T	30 ml	cooked vegetable, chopped[1]	1 c	240 ml
milk	¾ c	180 ml	eggs, separated	2	2
flour	2 T	30 ml			

1. Set oven at 325°F (165°C).
2. Melt fat. Stir in flour. Gradually stir in milk and cook, stirring constantly, until thickened. Add vegetable, remove from heat.
3. In a small bowl, beat egg yolks slightly.
4. Slowly add some of the vegetable mixture to yolks. Slowly pour warmed yolk mixture back into vegetable mixture, stirring constantly. Set aside to cool slightly.
5. In larger bowl, beat egg whites just until stiff. Gently fold vegetable–yolk mixture into egg whites; spoon into four 4-oz (120-ml) ungreased baking dishes.
6. Bake 25 to 30 minutes, or until puffy, golden brown and custard has set. (3 to 4 servings)

[1]Asparagus, spinach, broccoli, and peas may be used.

SUMMARY QUESTIONS—EGGS AND EGG PRODUCTS

1. Relating to egg quality:
   a. List three uses for which older eggs are satisfactory

   b. List three products for which high egg quality is essential

2. What causes the egg white to become thinner with age? Does this thinning affect the thickening power of eggs? The foaming power of eggs? Explain.

3. How does an increase in the following ingredients of a *soft* custard affect the *coagulation temperature*?
   Soft custard basic recipe: 2 c (480 ml) milk, 2 large eggs, ¼ c (60 ml) sugar

Added Ingredient	Change in Coagulation Temperature	Explanation
+ 2 T (30 ml) sugar		
+ ⅓ c (80 ml) milk		
+ 1 egg		

4. Outline briefly the *major* steps in the procedure for preparing a product that uses both starch and eggs as thickening agents.

5. Explain the technique used whereby eggs are successfully incorporated into a hot starch mixture.

6. What sanitary problems might occur in the following situations:
   a. Use of cracked egg for egg nog

   b. Storing dry powdered egg

c. Reconstituted powdered egg left unrefrigerated

d. Unpasteurized frozen eggs—whole, yolk, or white

e. Use of raw egg in Caesar Salad dressing.

f. Combining beaten egg white with fruit purée for a low-calorie dessert

7. Regarding egg substitutes used in laboratory:
   a. Are they all cholesterol free?

   b. What is the composition and what additives are used in the cholesterol free egg product?

   c. Identify how egg substitutes might be used in various food products

   d. Compare the cost and nutritive value of these products relative to fresh eggs

8. Investigate the current price of eggs per dozen. Which size is the "best buy"?

Size	Weight Per Dozen	Cost Per Dozen	Cost Per Egg
Extra large	27 oz (765 g)		
Large	24 oz (680 g)		
Medium	21 oz (595 g)		

Best buy:

156 Food Principles

9. Record the nutritive value of one serving of the following foods, using recipes where provided. Circle those foods that contribute more than 10% RDA of a nutrient.

	Energy (kcal)	Protein (g)	Calcium (mg)	Iron (mg)	Total vitamin A Activity (IU)	Thiamin (mg)	Riboflavin (mg)
Egg, 1 medium							
Baked custard							
Soufflé, cheese							
Pudding, vanilla							
R.D.A. (20-yr-old) Male							
Female							

G. Milk and Milk Products

OBJECTIVES

To recognize the variety of foods made from milk
To appraise milk product variations such as lowfat, fat free, low sodium cheeses and cheese products
To observe and describe reasons for coagulation of milk protein by several methods
To relate methods of coagulation to preparation and characteristics of several milk products
To observe, describe and relate the effect of heat on natural and processed cheese
To appraise nutritive, palatability, sanitary and economic characteristics of milk products

REFERENCES

Appendix G-I

ASSIGNED READINGS

TERMS

Listeriosis
Campylobacteriosis
Coagulation
Clot
Curd
Curdled
Casein

Whey
Enzyme
Substrate
Pasteurization
Homogenization
"Light" cheese
Low sodium cheese

Natural cheese
Processed cheese
Flavored cheese
Cold pack
Cheese spread
Imitation cheese

EXERCISE 1: COMPARISON OF MILK AND NON-DAIRY PRODUCTS

PROCEDURE

1. Sample various milk or nondairy products, e.g., lowfat, low-salt, nonfat, nonfat dry milk (NDM), lactose-reduced, imitation products, and yogurt.
2. Evaluate the palatability characteristic of each product (e.g., taste, consistency, acceptability).
3. Record the cost per serving and major nutrients supplied.
4. Summarize conclusions about the characteristics of the products, noting which may be considered the "best buy."

Sample Product	Palatability	Cost/Serving	Major Nutrients
1.			
2.			
3.			
4.			
5.			
6.			
7.			
8.			

Conclusions:

1. Delineate the percent composition of whole milk:
   Water: Carbohydrate: Fat: Protein: Minerals:

2. How does the percent of fat in whole milk compare with that of various other fluid milks?

3. What is filled milk? Imitation milk?

4. Which of the sampled products contain no dairy products?

EXERCISE 2: COAGULATION OF MILK PROTEIN

A. Addition of Acid

PROCEDURE

1. Bring 1 pint (480 ml) of whole milk to a boil.
2. Add 1 T (15 ml) lemon juice to the hot milk.
3. Bring the milk to a boil again.
4. Strain the mixture through a double thickness of cheese cloth.
5. Carefully squeeze out excess water. Refrigerate the cheese. Yield: ½ c (120 ml)

1. Describe the appearance of the milk-acid mixture after it is boiled.

2. What is the composition of the soft precipitate (soft cheese) and the whey?

B. Acid Produced by Bacteria (Yogurt)

PROCEDURE

1. Heat 1 quart (or liter) of milk to boiling. Pour into glass container.
2. Allow milk to cool to 111°F (44°C).
3. Stir 2 T (30 ml) plain commercial yogurt into the milk.
4. Cover and leave the mixture undisturbed in a warm location 111°F (44°C), for 3 to 5 hours until set.
5. Refrigerate. Yield: 1 qt (950 ml)

1. Describe the final product. How did it differ from the product made by addition of acid to milk in Part A?

2. In making yogurt, why is milk cooled to 111°F (44°C) before the bacterial culture is added?

3. How do the bacteria cause the coagulation of casein?

C. ENZYME ACTION (RENNIN)

PROCEDURE

1. Warm 2 c (480 ml) milk slowly to lukewarm 111°F (44°C); remove from heat immediately.
2. Empty 1 package of rennin pudding into milk, stirring until dissolved, not more than 1 minute.
3. Immediately pour mixture into custard cups and leave undisturbed for 10 minutes. Refrigerate. (4 servings)

1. Describe the final product.

2. What is the source of the enzyme rennin?

3. Identify the specific factors that are necessary for the optimum functioning of rennin.

4. How does the nutritive value of the product of rennin coagulation differ from that obtained when acid is used to coagulate casein? Why?

5. What other food products are made by rennin coagulation?

EXERCISE 3: COMBINING ACID FOODS WITH MILK

PROCEDURE

1. Using Alkacid test paper, test the pH of milk and tomato juice.
2. Combine milk and tomato juice in the following ways:
    a. Add ½ c (120 ml) cold tomato to ½ c (120 ml) hot milk.
    b. Add ½ c (120 ml) cold tomato to ½ c (120 ml) hot thickened[1] milk.
    c. Add ½ c (120 ml) cold tomato mixed with ⅛ t (.63 ml) soda to ½ c (120 ml) hot milk.
    d. Add ½ c (120 ml) hot thickened[1] tomato to ½ c (120 ml) cold milk.
    e. Add ½ c (120 ml) hot tomato to ½ c (120 ml) hot thickened milk.
3. Record the pH and evaluate the appearance of each mixture when first mixed.
4. Remove a sample of each mixture when the simmering temperature is reached. Evaluate the appearance.
5. Partially cover and cook mixture over low heat for 15 minutes longer. Evaluate the appearance.
6. Based on experiments, draw conclusions regarding the best method for combining an acid ingredient with milk.

[1]Where thickened tomato or milk product is required, use 1 T fat (15 ml) and 1 T (15 ml) flour.

pH milk: ___ pH tomato: ___	Initial pH of Mixture	Appearance		
		Initial Mixture	At Simmering	After 15 minutes
Cold tomato to hot milk				
Cold tomato to hot thickened milk				
Cold tomato plus soda to hot milk				
Hot thickened tomato to cold milk				
Hot tomato to hot milk				

Conclusions:

PALATABILITY TERMS—CHEESE									
FLAVOR				TEXTURE-CONSISTENCY					
Sharp Strong	Bland Acid	Salty Mild	Sour Sweet		Soft Creamy	Semisoft Curd	Crumbly Granular	Moist Dry	Firm Hard

EXERCISE 4: COMPARISON OF CHEESE PRODUCTS

PROCEDURE

1. Sample cheese products on display.
2. Complete table, noting palatability characteristics, uses and comparative cost.

Natural	Name	Palatability	Uses	Cost/lb (454 g)
Very hard	Parmesan			
Hard Ripened by bacteria (no eyes)	Cheddar			
	Edam–Gouda			
	Provolone			
Ripened by bacteria (with eyes)	Swiss			
Semisoft Ripened by bacteria	Muenster			
	Monterey Jack			
Ripened by blue mold (interior)	Roquefort			
	Blue			
Soft Ripened	Camembert			
Unripened	Cottage			
	Cream			
	Neufchatel (US)			
	Ricotta (Whey)			

Milk and Milk Products 163

Natural	Name	Palatability	Uses	Cost/lb (454 g)
Cheese Blends	Cold Pack			
Processed				
Low-salt				
Reduced-fat				

Courtesy: SYSCO ® Incorporated

EXERCISE 5: EFFECT OF HEAT ON NATURAL AND PROCESSED CHEESE

PROCEDURE

1. Preheat oven to 350°F (175°C).
2. Place 2 slices of bread on a baking sheet.
3. On one slice, place 3 T (45 ml) grated natural cheese; on the other slice, place 3 T (45 ml) grated processed cheese. Cut each slice in half.
4. Place baking sheet on upper shelf of oven. Remove one-half slice of each sample as soon as the cheese melts (3 to 5 minutes). Describe the appearance and texture.
5. Bake the remaining halves for an additional 5 minutes. Describe the appearance and texture.

	Appearance	Texture
Natural cheese 5 minutes		
10 minutes		
Processed cheese 5 minutes		
10 minutes		

Conclusions:

PALATABILITY TERMS—COOKED CHEESE

APPEARANCE	TEXTURE–CONSISTENCY		
Homogeneous	Smooth	Tender	Stringy
Separated	Uniform	Tough	Elastic
Curdled			

---

1. How does the composition of processed and natural cheese differ?

2. Compare the effects of heat on the cooked samples of natural and processed cheese. Explain any differences.

3. What effect does extended cooking at high temperatures have on cheese?

SUMMARY QUESTIONS—MILK AND MILK PRODUCTS

1. In general, what are the most satisfactory methods of preventing coagulation of milk casein in recipes using vegetables?

2. Discuss the use of soda to prevent curdling of vegetable milk combinations. Provide an example of when soda might be used.

3. In using a natural instead of processed cheese in a cheese sauce recipe, what special cooking techniques should be used to insure a smooth sauce?

4. In the preparation of pizza, a hot oven temperature (450° to 500°F, 230° to 260°C) is used. How is the adverse effect of high heat on cheese protein minimized?

5. Did coagulation of casein play a role in the thickening of egg custards (egg and milk mixtures)? In the curdling of the overheated custard? Explain.

6. Based on laboratory experiments and readings, list the components in the following products that might cause coagulation of casein. How could coagulation be prevented?

Product	Factor	Prevention
Scalloped potatoes		
Cream of asparagus soup		
Ham slices baked in milk		
Milk–fruit juice beverage		

166 FOOD PRINCIPLES

7. What practical suggestions can be offered to an individual who likes milk and milk products but needs to restrict calories?

8. Consult appropriate references and list several cheeses that are relatively low in calories and several that are low in sodium.

9. Complete the nutritive value chart.

Food	Energy (kcal)	Protein (g)	Fat (g)	Calcium (mg)	Vitamin A Act. (IU)	Riboflavin (mg)
Whole milk (1 c, 240 ml)						
2% milk (1 c, 240 ml)						
.5% milk (1 c, 240 ml)						
Nonfat milk (1 c, 240 ml)						
Sour cream (1 c, 240 ml)						
Lowfat (1 c, 240 ml)						
Yogurt, low-fat (1 c, 240 ml)						
Flavored (1 c, 240 ml)						
Cottage cheese (1 c, 240 ml)						
Lowfat (1 c, 240 ml)						
Nonfat (1 c, 240 ml)						
Cheddar cheese (1 oz, 28 g)						
Reduced-fat (1 oz, 28 g)						
Processed cheese (1 oz, 28 g)						
Imitation cheese (1 oz, 28 g)						

Milk and Milk Products 167

10. Account for differences in calcium content of cheddar cheese and cottage cheese.

11. In terms of sanitary quality, what problems may occur in the use of milk products? Explain.

12. What is lactose? How is lactose reduced in milks to provide more digestible milk for lactose-intolerant people?

13. Compare the cost of 10 g of protein from whole milk, cottage cheese, cheddar cheese, ground beef, and peanut butter.

14. What suggestions could be offered to someone on a low income about how to use nonfat dry milk? Consider especially the problem encountered by an individual who finds the taste unacceptable.

15. How does yogurt cheese differ from that made in Exercise 2A? Yogurt cheese is made as follows: 1 qt (.95 l) plain, low-fat yogurt is poured into a filter or cheese cloth lined sieve, which is placed over a large bowl. Cover and refrigerate about 8 hours or overnight. The yield is 2 cups (480 ml) of a creamy spread or slightly less if a firmer cheese is desired. Season with herbs.

H. Batters and Doughs

OBJECTIVES

To describe the function of ingredients in a variety of products made from batters and doughs
To compare the gluten potential of flours made from wheat, corn, rye and soy
To assess the effectiveness of different leavening agents and relate these to palatability characteristics
To delineate relationships of kind and proportion of ingredients to final product characteristics
To evaluate the effect of manipulation on gluten development in a variety of batters and doughs
To relate gluten development to palatability characteristics of products made from various batters and doughs
To distinguish palatability characteristics such as flakiness, tenderness and grain in batters and doughs
To compare subjective and objective measurements in assessing palatability of pastry
To evaluate nutritive value of different grains

REFERENCES

ASSIGNED READINGS

TERMS

All purpose flour	Conventional method	Grain
Whole wheat flour	Dump method	Tunnels
Rye flour	Hydration	Crumbs
Cornmeal	Knead	Cell walls
Soy flour	Cream	Cell size
Gluten potential	Fold	Saturated fat
Gluten development	Cohesive	Unsaturated fat
Leavening	Elastic	Hydrogenated fat
Fermentation	Flakiness	Plastic fat
Muffin method	Tenderness	Emulsifier
Pastry method	Oven spring	

BATTERS AND DOUGHS

WHOLE WHEAT BREAD

Source: Division of Nutritional Sciences, New York State College of Human Ecology at Cornell

EXERCISE 1: MEASUREMENT OF FLOUR

PROCEDURE

1. Weigh 1 c (240 ml) all-purpose flour as directed on chart. Sift flour. Gently scoop sifted flour into measuring cup without packing. Level off with straight-edge knife and weigh.
2. Compare results with classmates and account for differences.

	Weight (g)	Class Range and Average (g)
1 c (240 ml) flour, *unsifted*		
1 c (240 ml) flour, *sifted*		

Why are there differences in weight between unsifted and sifted flour?

EXERCISE 2: STRUCTURAL PROPERTIES OF WHEAT FLOUR

PROCEDURE

Either prepare or observe a demonstration of making gluten balls as follows:

1. Mix 1 c (240 ml) all-purpose flour with approximately ¼ c (60 ml) water until all of the water is absorbed.
2. Knead dough 10 to 15 minutes until it is cohesive and elastic.
3. Place dough in a bowl filled with cold water; squeeze the dough to work out the starch. Repeat the process with fresh water until the bowl water is clear.
4. Press water from the dough. Bake dough ball at 400°F (205°C) for about 30 minutes, until firm.

1. Why is cold water used to remove the starch?

2. As flour and cold water are mixed to make a dough, what is happening to the *starch* component of the flour?

3. As the dough is kneaded, what is happening to the *protein* in the flour?

4. How does the baked gluten ball differ from the unbaked?

5. What is the source of leavening in gluten balls?

6. What is meant by the gluten potential of a flour?

7. Why are gluten balls from cake flour, all-purpose flour, and bread flour different in size?

GLUTEN BALLS, UNBAKED AND BAKED. (LEFT TO RIGHT) CAKE FLOUR, ALL-PURPOSE FLOUR, AND BREAD FLOUR

Courtesy: Wheat Flour Institute

EXERCISE 3: CHEMICAL LEAVENING AGENTS

A. Ingredients of Baking Powders

PROCEDURE

1. Inspect labels on several baking powder cans.
2. Complete the following chart.

Baking Powder	Alkali	Acid	Teaspoons/Cup Flour (ml/240 ml)
1.			
2.			
3.			

B. Comparison of Speed of Reaction

PROCEDURE

1. Mix 1 t (5 ml) specified leavening with liquid (1 T [15 ml]) as directed.
2. Observe speed of reaction.
3. Summarize conclusions about speed of reaction and ingredients.

	Liquid	Relative Speed of Gas Production
Baking soda	1 T cold water	
Baking soda	1 T hot water	
Baking soda	1 T vinegar	
Baking soda + ¼ t (1.25 ml) cream of tartar	1 T cold water	
Tartrate baking powder	1 T cold water	
Double-acting baking powder	1 T cold water	
Double-acting baking powder	1 T hot water	

Conclusions:

EXERCISE 4: FACTORS AFFECTING THE LEAVENING POWER OF YEAST

PROCEDURE

1. Prepare STANDARD mixture as follows for each of three small custard cups:
    - 1 package dry yeast
    - 2 T (30 ml) water, room temperature
    - 1 T (15 ml) flour
    - ¼ t (1.25 ml) sugar
2. Add the sugar and salt variables to the STANDARD mixture as directed. Stir and allow all mixtures to react for 25 minutes.
3. Observe rate of gas production and summarize results.

Variable	Rate of Gas Production
STANDARD	
STANDARD + 2 T (30 ml) sugar	
STANDARD + 1 t (5 ml) salt	

QUESTIONS—LEAVENING AGENTS

1. List some acid foods that are commonly used with baking soda.

2. To what components of baking powder do the terms "single-acting" and "double-acting" refer? Why are the terms accurate?

3. What is the role of starch in baking powders?

4. Account for the "soapy" taste in products that have excess soda.

5. Why are soda-acid leavened products often extremely tender?

6. What environmental factors must be present to insure adequate growth for yeast?

7. How does the leavening action of yeast differ from that of baking powder?

OATMEAL MUFFINS

Source: Cornell University, Cooperative Extension

EXERCISE 5: DROP BATTER, MUFFINS[1]

Basic Muffin Recipe

sifted all-purpose flour	1 c	240 ml	egg	1	1
sugar	2 T	30 ml	milk	½ c	120 ml
double-acting baking powder	1½ t	7.5 ml	oil	2 T	30 ml

1. Grease 6 muffin cups (or use paper liners) and set oven at 425°F (220°C).
2. Sift the dry ingredients into a mixing bowl.
3. Beat the egg slightly, add milk and oil.
4. Make a well in the dry ingredients. Add the liquid ingredients and stir until the dry ingredients are just moistened (about 16 stirs). The batter will be lumpy.
5. Fill the greased muffin cups half-full.
6. Bake the muffins about 20 minutes, until golden brown.

[1]Refer to Microwave Cooking chapter for microwave recipes.

A. Effect of Manipulation

PROCEDURE

1. Follow BASIC MUFFIN recipe, mixing 16 stirs or until dry ingredients are *just moistened*.
   a. Place 2 portions (about ⅓ c, 80 ml) of batter in muffin pan.
   b. Stir remaining batter 5 additional strokes and remove 2 portions.
   c. Stir remaining batter 25 additional strokes and remove 2 portions.
2. Evaluate products on the chart below.
3. Summarize effects of manipulation on palatability characteristics of muffins.

Characteristic	Amount of Manipulation		
	Mix until Moistened	Additional 5 Strokes	Additional 25 Strokes
Color			
Shape			
Volume			
Tenderness			
Grain (cell size, tunneling, etc.)			
Extent of gluten development			

B. Effect of Different Grains

PROCEDURE

1. Follow the BASIC MUFFIN Recipe substituting other flours[1], as directed. (50% = ½ All-purpose)
2. Evaluate all products on the following chart. Summarize results.

Flour	Volume	Texture	Tenderness of Crumb	Extent of Gluten Development
Cornmeal 100%				
50%				
Whole wheat 100%				
50%				
Rye 100%				
50%				
Soy 50%				
25%				
Other				

EFFECT OF MANIPULATION ON MUFFINS. (LEFT TO RIGHT) UNDERMIXED, MIXED TO MOISTEN, AND OVERMIXED.
Source: Division of Nutritional Sciences, New York State College of Human Ecology at Cornell.

[1] Suggested as partial substitution for all purpose flour: oat bran; wheat bran; wheat germ; triticale.

CORRECTLY MIXED BATTER, (LEFT) MUFFIN HAS ROUNDED TOP, EVEN TEXTURE; (RIGHT) OVERMIXED BATTER. MUFFIN HAS TUNNELS AND PEAK.

Source: Division of Nutritional Sciences, New York State College of Human Ecology at Cornell.

PALATABILITY STANDARD—MUFFINS

APPEARANCE	TEXTURE OF CRUMB	TENDERNESS
Volume: double unbaked Top: uneven, pebbled 　　　slightly rounded 　　　golden brown	Uniform Air cell: medium-coarse Cell walls: medium thick	Breaks easily Soft in mouth

QUESTIONS—MUFFINS

1. What is the ratio of liquid to flour in a muffin recipe? How does this ratio affect gluten development?

2. What are the sources of leavening in muffins?

3. Why do most muffin recipes specify liquid fat (oil or melted solid)? How does fat function in a muffin batter?

4. What causes tunnels? Were tunnels prevalent in muffins made with the corn meal and whole wheat flour? Explain.

5. Why does wheat flour have a high gluten potential? Why are corn and rye low?

6. How is a desirable structure be obtained if low gluten potential flour (rye, cornmeal) is desired in a muffin recipe?

BATTERS AND DOUGHS

EXERCISE 6: SOFT DOUGH, BISCUITS[1]

Basic Biscuit Recipe

sifted all-purpose flour	2 c	480 ml	fat, hydrogenated	¼ c	60 ml
double acting baking powder	1 T	15 ml	milk	⅔ c	160 ml

1. Set oven at 425°F (220°C).
2. Sift the dry ingredients into a mixing bowl.
3. Cut the fat into the dry ingredients, using a pastry blender or two knives, one in each hand. Continue cutting until no fat particles are larger than peas.
4. Add the milk and mix vigorously with a fork until the dough is stiff (about 25 times), cutting through the center of the dough with the fork several times.
5. Knead 10 times, roll to 1-inch (2.54-cm) thickness.
6. Cut and bake 12 minutes or until brown.

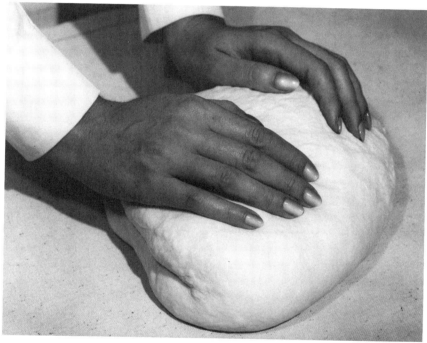

KNEADING DOUGH
Courtesy: Wheat Flour Institute

PALATABILITY STANDARD—BISCUITS

APPEARANCE	TEXTURE	TENDERNESS
Volume: doubled	Uniform	Crust: crisp, easy to break
Top: golden brown flat, circular	Air cell size: small	Crumb: soft to touch, moist
Sides: straight	Cell walls: thin	
	Flaky layers	

[1] Refer to Microwave Cooking chapter for microwave recipes.

A. Effect of Manipulation

PROCEDURE

1. Follow BASIC BISCUIT Recipe through step 4.
2. Divide dough into 3 portions.
3. Lightly flour board and rolling pin.
4. Manipulate dough, kneading as directed. Roll to 1 inch (2.54 cm) thick and bake for 12 minutes.
5. Evaluate products and summarize the effect of manipulation on palatability of biscuits.
6. Save biscuits from manipulation variation b to use as STANDARD for Part B.

Palatability	Amount of Manipulation		
	a. No kneading	b. Kneaded 10 Times	c. Kneaded 30 Times
Appearance			
Volume			
Shape			
Texture			
Cell size			
Flakiness			
Tenderness			
Extent of gluten development			

Conclusions:

EFFECT OF MANIPULATION ON BISCUITS. (LEFT TO RIGHT) NO KNEADING, KNEADED 10 TIMES, AND KNEADED EXTENSIVELY.

Source: Division of Nutritional Sciences, New York State College of Human Ecology at Cornell.

B. Substituting Soda Acid for Baking Powder

PROCEDURE

1. Follow BASIC BISCUIT Recipe, substituting ⅔ c (160 ml) buttermilk for regular milk and using ½ t (2.5 ml) baking soda and 2 t (10 ml) double acting baking powder as leavening.
2. Mix as directed and knead 10 times.
3. Evaluate products, comparing soda-acid biscuits with STANDARD baking powder biscuits (in Part A, variation b). Record observations.

Palatability	Standard-Baking Powder	Soda-Acid
Appearance Volume		
Shape		
Color of crust		
Texture Cell size		
Flakiness		
Tenderness		

QUESTIONS—BISCUITS

1. What is the ratio of liquid to flour in a biscuit recipe?

2. How does this ratio affect the development of gluten?

3. What are the sources of leavening in the biscuits?

4. How does fat function in a biscuit recipe?

5. How would the state of the fat (solid or liquid) affect the texture of a biscuit?

6. Based on the experiments, how do the palatability characteristics of a soda acid biscuit differ from a biscuit leavened by baking powder?

7. In substituting soda for baking powder, what are the proportions used? What is the amount of soda used to neutralize 1 c (240 ml) buttermilk or sour milk?

EXERCISE 7: PANCAKES, POPOVERS, CREAM PUFFS

A. Effect of Manipulation on Gluten Development in Pancakes

Basic Pancake Recipe[1]

sifted all purpose flour	1¼ c	300 ml	egg	1	1
baking powder	1¾ t	9 ml	milk	1 c	240 ml
sugar	2 T	30 ml	oil	2 T	30 ml

1. Sift dry ingredients into medium bowl.
2. Beat wet ingredients and add to dry ingredients.
3. Stir quickly only until ingredients are combined; batter will be somewhat lumpy.
4. Bake on hot griddle or heavy skillet until bubbles form on surface and edges become dry. Turn, cook approximately 2 minutes until golden brown.

Procedure

1. Follow BASIC RECIPE through step 2, but mix as follows:
   a. Stir liquid and dry ingredients only until moistened. Remove ½ c (120 ml) batter and bake 2 pancakes.
   b. Stir remaining batter an additional 25 strokes and bake 2 pancakes.
2. Evaluate products and summarize results.

---

[1] As a variation, use ½ c (120 ml) whole wheat flour and ¾ c (180 ml) all-purpose flour.
Picture Courtesy: USDA

BATTERS AND DOUGHS

Palatability	a. Until Moistened	b. Additional 25 Strokes
Appearance		
Texture		
Tenderness		
Extent of gluten development		

Conclusions:

PALATABILITY STANDARD—PANCAKE

APPEARANCE
Volume: double unbaked
Shape: regular
Color: evenly browned

TEXTURE
Uniform, even
Air cell size: medium-fine
Cell walls: medium

TENDERNESS
Crust: easy to cut
Crumb: light, moist, not gummy

B. Effect of Manipulation on Gluten Development in Popovers

Basic Popover Recipe

flour, sifted	½ c	120 ml	milk	½ c	120 ml
egg	1	1			

1. Set oven at 425°F (220°C) and *lightly grease* four 5-oz (150-ml) custard cups.
2. Sift flour into small bowl. Add egg and milk.
3. Beat with electric or rotary beater until smooth.
4. Pour into custard cups, place on baking sheet.
5. Bake for 40 to 45 minutes, or until golden brown, reducing oven temperature to 375°F (190°C) after 20 minutes.

PROCEDURE

1. Follow BASIC POPOVER recipe through step 3, then proceed as follows:
   a. Place half of batter into 2 greased custard cups. Bake as directed.
   b. Beat the remaining batter an additional minute. Pour this batter into 2 greased custard cups. Bake as directed.

2. Compare products and summarize results.

	Beaten Until Smooth	Beaten Additional 1 Minute
Appearance		
Texture		
Tenderness		
Crust		
Crumb		

Conclusions:

PALATABILITY STANDARD—POPOVER

APPEARANCE

Volume: double or triple unbaked
Color: medium brown shiny crust

TEXTURE

Crust: crisp
Inside: hollow, moist, but not soggy light, airy

C. Cream Puffs

Basic Cream Puff Recipe

| water | ½ c | 120 ml | flour, sifted | ½ c | 120 ml |
| margarine | ¼ c | 60 ml | eggs, medium | 2 | 2 |

1. Set oven at 450 °F (230°C) and *lightly grease* a baking sheet.
2. Place water and fat in medium saucepan and bring to a boil to melt fat.
3. Immediately add all the flour and stir vigorously until the batter is smooth and forms a ball. Remove pan from the heat.
4. Add eggs, one at a time, beating well after each addition.
5. Drop batter into 4 mounds onto baking sheet and bake for 15 minutes; reduce oven temperature to 350°F (175°C). Bake 20 minutes longer or until puffs are lightly brown and firm.

PROCEDURE

1. Prepare cream puffs according to recipe.
2. Evaluate palatability of final product; account for any differences from Palatability Standard.
   - Appearance:
   - Texture:
   - Tenderness:

PALATABILITY STANDARD—CREAM PUFF		
APPEARANCE	**TEXTURE**	**TENDERNESS**
Volume: double unbaked Shape: rounded Color: golden brown	Hollow center, 1 large hole	Crust: crisp, tender Interior: slightly moist

QUESTIONS—PANCAKES, POPOVERS, AND CREAM PUFFS

1. What is the proportion of liquid to flour in pancakes and popovers? What effect does this proportion have on the development of gluten?

2. How are pancakes leavened? How are popovers leavened?

3. Predict the effect on palatability if pancakes are turned after bubbles burst. What are the bubbles?

4. Why is a high initial oven temperature essential to the leavening of popovers and cream puffs?

5. In addition to contributing to structure, how does the egg function in the cream puff dough?

EXERCISE 8: STIFF DOUGH—YEAST BREAD/ROLLS

PROCEDURE

1. Prepare yeast rolls according to the recipe below.
2. Evaluate final product. If the product differs from Palatability Standard. Explain.

Yeast Rolls

water, warm	⅔ c	160 ml	salt[2]	1 t	5 ml
dry yeast[1]	½–1 pkg	½–1	flour[3]	2½–3 c	600–720 ml
sugar	2 T	30 ml	egg (optional)	1	1
nonfat dry milk solids	⅓ c	80 ml	vegetable oil	2 T	30 ml

1. Place warm water (105° to 115 °F, 40° to 46°C) in a large mixing bowl.
2. Stir the dry yeast and sugar into water.
3. Add milk and salt to yeast water mixture.
4. Stir in the flour, 1 c (240 ml) at a time, alternating with the eggs and oil. Beat well after each addition.
5. Add enough of the remaining flour to make a soft dough.
6. Turn dough onto a well floured board and knead until smooth and elastic. The dough will spring back when touched.
7. Place dough in slightly oiled bowl; oil the top of the dough, cover tightly.
8. Allow to double in volume at room temperature. Dough has risen sufficiently if pressed finger marks remain in the dough.
9. Punch dough down and knead until smooth and elastic on an unfloured (or lightly floured) board.
10. Shape the dough into desired type of rolls (see Shaping Yeast Dough illustration).
11. Place the rolls on a lightly greased pan and brush with oil. Set oven at 400°F (205°C).
12. Allow rolls to double in volume (about 45 minutes, on counter). Bake for 10 to 15 minutes, or until browned. (20 medium rolls)

[1] If laboratory time is short, use 1 package yeast
[2] Salt is a required ingredient because it regulates the growth of yeast.
[3] Up to ½ of the flour can be whole wheat.

TESTING RISEN DOUGH

PUNCHING DOWN DOUGH

Courtesy: Wheat Flour Institute

SHAPING YEAST DOUGH
Courtesy: USDA

PALATABILITY STANDARD—YEAST ROLLS

APPEARANCE	TEXTURE	TENDERNESS	TASTE
Symmetrical Crust: golden smooth Volume: doubled	Uniform Air cells: medium-fine Cell walls: thin	Crust: thin, easy to cut Crumb: silky, moist	Fresh Not yeasty Not flat

EVALUATION OF YEAST ROLLS

	Reasons for Variation from Standard Product (if applicable)
Appearance	
Shape	
Volume	
Texture	
Uniformity	
Cell size	
Cell walls	
Tenderness	
Crust	
Crumb	
Taste	

YEAST BREAD. NOTE EXCELLENT VOLUME, UNIFORM TEXTURE AND MEDIUM-SIZE AIR CELLS.

Courtesy: USDA

QUESTIONS—YEAST ROLLS

1. What is the ratio of liquid to flour in yeast dough? How does this ratio affect gluten development?

2. List the products of yeast dough fermentation and state how each affects the quality of the baked product.

3. Explain why extensive gluten development is advantageous in yeast-leavened products.

4. List the ingredients that are *essential* for breadmaking. Explain.

5. Is it necessary to scald milk(s) for addition in a bread recipe? At what temperature should fluid milk be, before combining it with yeast? Why? How may milk temperature be determined without use of a thermometer?

6. What are the effects of allowing yeast to rise too much? Too little?

7. What is meant by "oven spring"?

8. Predict the relative volume of a sweet yeast dough compared to the yeast roll recipe used in laboratory.

9. Identify what happens to the ingredients in yeast rolls during the following processes:

	Flour	Liquid	Yeast	Sugar
Mixing ingredients				
Kneading				
Rising				
Baking				

EXERCISE 9: SHORTENED CAKES

A. Effect of Manipulation and Type of Shortening on Cake Texture

PROCEDURE

1. Prepare a cake using assigned method of mixing and assigned shortening.
2. Evaluate batter characteristics and palatability characteristics of all variables. Record observations.

PALATABILITY STANDARD—CAKE		
APPEARANCE	TEXTURE OF CRUMB	MOUTHFEEL
Volume: double unbaked Top: slightly rounded golden brown	Uniform Air cell size: small–medium Cell Walls: thin	Slightly moist Velvety Light

Shortened Cake[1]

shortening	⅓ c	80 ml	sifted cake flour	1 c	240 ml
vanilla	½ t	2.5 ml	or ⅞ c (210 ml) all purpose flour		
sugar	⅔ c	160 ml	double acting baking powder	1 t	5 ml
eggs	1½	1½	milk	⅓ c	80 ml

CONVENTIONAL METHOD OF MIXING

1. Set oven at 350°F (175°C) and grease a 6 × 6 inch (15 × 15 cm) or 7 × 11 inch (18 × 28 cm) pans.
2. Add vanilla to assigned shortening and cream until soft.
3. Gradually add sugar to softened fat and cream until light and fluffy using hand mixer set on medium speed. (Mixing by hand: add 2 T (30 ml) of sugar at a time and beat 100 strokes after each addition).
4. Add unbeaten eggs, mixing until blended (150 strokes by hand).
5. Sift dry ingredients. Add dry ingredients alternately with milk, beginning and ending with flour. Mix just until a smooth batter is formed.
6. Pour batter into prepared pan. Bake for 30 minutes, or until a toothpick inserted into cake comes out clean.
7. Set cake on rack and cool 10 minutes before removing the cake from the pan.

CONVENTIONAL CAKE. NOTE EXCELLENT VOLUME, UNIFORM TEXTURE, AND THIN CELLS WALLS

Courtesy: General Mills

DUMP METHOD OF MIXING

1. Sift dry ingredients into a large bowl.
2. Add wet ingredients.
3. Beat batter with hand mixer set on medium speed, for 2 minutes, cleaning sides of bowl. Beat for 2 additional minutes.
4. Pour batter into prepared pan and proceed as directed for conventional cake.

[1] Use butter, margarine or hydrogenated fat. May experiment with a fat-free spread.

EVALUATION OF CAKES

Shortening	Appearance	Texture	Mouthfeel
Butter Conventional			
Dump			
Margarine Conventional			
Dump			
Hydrogenated fat Conventional			
Dump			

Conclusions:

QUESTIONS—CAKES

1. In the conventional method of mixing, what is the purpose of creaming?

2. What is the function of eggs in shortened cakes?

3. Which assigned shortening contained an emulsifier? What is an emulsifier, and how does it act to improve the cake quality?

4. Why are differing amounts of flour (all-purpose and cake) used for the cake recipe?

5. In addition to fat, what other ingredients influence volume and texture of the cake?

EXERCISE 10: STIFF DOUGH—PASTRY[1,2]

A. Effect of Different Fat Plasticity on Palatability of Pastry

PROCEDURE

1. Follow BASIC PASTRY Recipe to prepare a one-crust pie shell using assigned shortening (at room [rm] or refrigerator temperature [ref]).
2. Evaluate all variables and record evaluations.
3. Summarize the relationship between type of shortening and subjective measurements of palatability.

Basic Pastry Recipe

flour, sifted	¾ c	180 ml	shortening (fat or oil)	¼ c	60 ml
water	1 T + 2 t	25 ml			

1. Set oven at 375°F (190°C).
2. Cut solid fat into the flour until the largest particles are about the size of peas.
3. Add the water and stir with a fork until the dough leaves the sides of the bowl and holds together.
4. Lightly dust the board or pastry cloth and a rolling pin with flour and roll the ball into a circle ⅛ inch (.32 cm) thick.
5. Fold the rolled pastry in half or roll it around the rolling pin, and place it in a 7″ (18-cm) pie plate. Gently unfold or unroll, ease crust into pie plate and crimp edge. Prick shell.
6. Place pie shell on *middle shelf* of oven and bake 10 to 12 minutes until lightly browned.

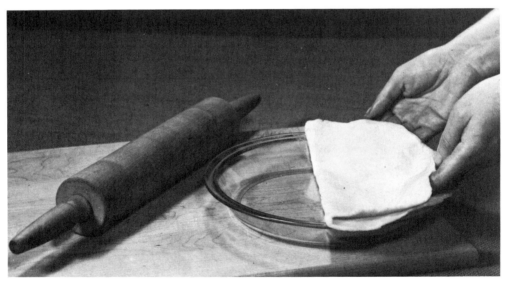

PLACING CRUST IN PIE PLATE.

Courtesy: USDA

[1] Refer to Microwave Cooking chapter for microwave recipes.
[2] Recipes yield a 7″ (18-cm) pie.

EVALUATION OF PASTRY[1]

Palatability Characteristics	Lard		Hydrogenated Fat		Butter		Margarine		Oil	
	rm	ref	rm	ref	rm	ref	rm	ref	rm	ref
External appearance Color uniform Pale Surface broken										
Internal Tenderness Very tender Fairly tender Crumbly Tough Mealy										
Flakiness Thin flakes Some flakes Thick flakes No flakes										
Flavor Pleasing Tasteless Displeasing										

Conclusions:

[1] May evaluate fat-free spread.

B. Effect of Different Fillings on Palatability of Bottom Crust

PROCEDURE

1. Prepare pastry. Use assigned shortening and follow BASIC PASTRY Recipe, doubling recipe if required for a two-crust pie.
2. Prepare assigned fillings (fresh fruit, cooked fruit, starch thickened, or custard).
3. Evaluate finished product (crust and filling).

Apple Filling (Fresh Fruit)

double BASIC PASTRY recipe					
cooking apples peeled and sliced	2–3	2–3	flour	1 T	15 ml
cinnamon, nutmeg or other spices	¼ t	1.25 ml	sugar (opt)	2 T	30 ml

1. Set oven at 425°F (220°C). Follow BASIC PASTRY directions through step 5.
2. Roll out top crust. Make small cuts for steam to escape.
3. Fill bottom crust with apple filling. Sprinkle with spices, flour, and sugar.
4. Moisten the edge of bottom crust with water. Place top crust on apples and press the crusts together to seal.
5. Trim surplus dough and crimp edge.
6. Place pie on *lower oven shelf*. Bake for 15 minutes.
7. Move pie to center or top rack, reduce oven temperature to 375°F (190°C) and bake 30 more minutes. (7-inch [18-cm] pie)

Cherry Pie (Starch Thickened)

double BASIC PASTRY recipe		
canned cherry pie filling, heated to boiling	1½ c	360 ml

1. Set oven at 425°F (220°C). Follow directions for BASIC PASTRY through step 5.
2. Place cherry pie filling in bottom crust, moisten edge, cover with top crust, seal and vent.
3. Place pie on lower oven shelf and bake for about 30 minutes, until crust is brown. (7-inch [18-cm] pie)

Chocolate Meringue Pie (Starch Thickened)

BASIC PASTRY recipe

sugar	½ c	120 ml	egg yolks	2	2
flour	3 T	45 ml	vanilla	½ t	2.5 ml
milk	1⅓ c	320 ml	egg whites	2	2
square chocolate, chopped	1	1	sugar	¼ c	60 ml

1. Set oven at 375°F (190°C). Follow BASIC PASTRY for a single crust recipe. Bake 15 minutes.
2. In saucepan, combine ½ the sugar and flour. Stir in milk and chocolate, cooking until the mixture thickens; add remaining ½ sugar.
3. Slowly add part of the hot mixture to the egg yolks (tempering). Return warmed egg to the chocolate mixture. Cook over low heat, stirring to coagulate the egg.
4. Remove the mixture from heat; add vanilla. Cool slightly. Pour filling into baked pie shell.
5. Set oven at 425°F (220°C). Beat egg whites until soft peaks develop.
6. Add the sugar, about 2 T (30 ml) at a time, beating after each addition.
7. Continue to beat the egg whites until stiff peaks develop.
8. Spread the meringue over the warm, but not hot filling, spreading to touch the crust.
9. Bake for 4 to 5 minutes or until lightly brown. (7-inch [18-cm] pie)

Quiche (Custard)

BASIC PASTRY recipe[1]

finely chopped, cooked vegetables[2]	1 c	240 ml	grated cheese	¼ c	60 ml
			milk	⅔ c	160 ml
minced onion	2 T	30 ml	eggs	2	2
oregano	¼ t	1.25 ml	nutmeg	¼ t	1.25 ml

1. Follow BASIC PASTRY Recipe, but bake only 5 minutes. Lower oven temperature to 350°F (175°C).
2. Layer vegetables, herbs and cheese in crust.
3. Beat milk and eggs and pour over vegetables. Sprinkle with nutmeg.
4. Place pie on *lower oven shelf*, and bake at 350°F (175°C) 30 minutes. If a knife inserted into the custard does not come out clean, move pie to center rack, lower heat to 325°F (165°C) and bake until custard is firm and knife comes out clean. Let stand 10 minutes before serving. (2 to 3 servings; 7-inch [18 cm] pie).

[1] Whole wheat or half whole wheat flour suggested.
[2] Suitable vegetables include: broccoli, asparagus, spinach, mushrooms.

PALATABILITY OF PIES

	Crust	Filling
Apple		
Cherry		
Quiche		
Chocolate Meringue		

VEGETABLE QUICHE

Source: Division of Nutritional Sciences, New York State College of Human Ecology at Cornell.

Delineate the general procedures followed in preparing and baking crusts for the four basic types of filling:

	Procedure With Crust
Fresh fruit	
Cooked fruit	
Starch thickened	
Custard	

QUESTIONS—PASTRY

1. Explain how "spreadibility" of a fat affects texture and tenderness of pastry. Which fat is most spreadable at room temperature? Refrigerated?

2. What is meant by "plastic" when describing a fat? Give an example of a plastic fat.

3. Discuss how hydrogenation changes the characteristics of a fat.

4. What is the cause of a "mealy" textured pastry?

5. In the marketplace, compare the nutrient label panels of several margarines.

    a. What information on the label helps the consumer select a margarine that is high in polyunsaturated fats?

    b. What additives are commonly used in the manufacture of margarines? List several and note the function.

6. a. What difficulty occurs when oil is used in the BASIC PASTRY method? Why?

    b. The following is a recipe designed to use oil as the fat in making pastry. Based on your experiences, explain why a tender (not mealy), somewhat flaky pastry can be obtained using this method.

    | flour, sifted | 1 c | 240 ml | oil | ¼ c | 60 ml |
    | salt | ½ t | 2.5 ml | milk, cold | 2 T | 30 ml |

    1. Sift flour and salt.
    2. Combine oil and milk, mixing well.
    3. Add liquids all at once to dry ingredients, and stir to form a moist ball.

7. Explain how the following factors affect gluten development in pastry:

Factor	Effect on Gluten	Explanation
Amount of fat		
Plasticity of fat		
Temperature of fat		
Amount of mixing of fat into flour		
Amount of liquid		
Amount of mixing of water into fat-flour		
Mixing ½ whole wheat, ½ all purpose flour		

SUMMARY QUESTIONS—BATTERS AND DOUGHS

1. Compare the nutritive value of the following grain products:

Grain	Energy (kcal)	Protein (g)	Calcium (mg)	Iron (mg)	Thiamin (mg)	Riboflavin (mg)	Niacin (mg)	Folate (mg)
Wheat flour, all-purpose, enriched								
Cake flour								
Whole wheat flour								
Cornmeal Enriched								
Unenriched								
Rye flour								
Soy flour								

2. Complete the following table regarding gluten potential.

Type of Flour	% Protein[1]	Gluten Potential	Uses
Wheat flour Whole wheat	13		
Hard wheat	12		
Soft wheat	9		
All-purpose	10		
Cake	7–8		
Cornmeal	7–9		
Rye flour	9–11		
Soy flour	34–47		

[1] Watt BK, Merrill AL. *Composition of Foods.* Agriculture Handbook, No. 8, USDA, 1963.

3. Analyze the following batters and doughs:

Product	Proportion Liquid/flour	Ease of Gluten Development	Description of Product With Overdeveloped Gluten
Biscuits			
Muffins			
Pancakes			
Popovers			

4. Indicate how the following variables can affect *gluten* development in a standard muffin and biscuit recipe:

Ingredient	Muffin	Biscuit
Decrease sugar		
Increase fat		
Increase liquid		
1 c (240 ml) cornmeal for 1 c (240 ml) flour		
1 c (240 ml) bread flour for 1 c (240 ml) all-purpose flour		

5. What other products besides muffins are prepared by the "muffin" method? What other products are prepared by the "pastry" method?

6. Explain why salt is considered an essential ingredient in yeast breads.

7. What are the functions of a liquid that are common to all batters and doughs?

8. What happens to the following ingredients during baking?

	Reaction During Baking
Flour	
Fat	
Milk	
Egg	
Baking powder (double-acting)	

9. List three principal leavening gases, and state the manner in which each may be incorporated into batters and doughs.

10. Could yeast be substituted for baking powder in a biscuit recipe? What changes in procedure would be necessary? Would characteristics of the end product be different? Explain.

11. If biscuits and yeast rolls were each kneaded for 10 minutes, would both products be equally palatable? Explain.

12. Regarding the characteristic of flakiness, describe:
    a. What is meant by flakiness in pastry or biscuits

    b. How ingredients are manipulated to obtain flakiness

    c. Why some gluten development is necessary for flakiness

    d. Why flakiness is not achieved by the muffin method

13. a. List several batter and dough convenience foods that are available to consumers.

    b. Compare cost and ingredients in three batter and dough convenience foods with those of similar products prepared from "scratch."

I. Fats and Oils

OBJECTIVES

To illustrate some factors that affect the formation and stability of food emulsions
To apply the concepts of food emulsions to a variety of food products
To evaluate the effects of various fats and oils, and fat replaced food products
To evaluate the palatability, cost and nutritive value of fat free and fat reduced products

REFERENCES

ASSIGNED READINGS

TERMS

Dispersed phase	French dressing	Sauté
Continuous phase	Mayonnaise	Fat free
Surface tension	Cooked dressing	Fat substitute
Emulsifier	Pan fry	Nonstick cooking spray
Lecithin		

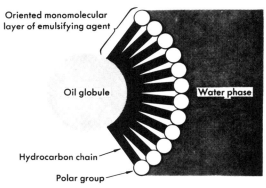

THE ORIENTATION OF EMULSIFYING AGENT IN AN OIL-IN-WATER EMULSION

Source: Hartman JR, 1977. Colloid Chemistry. Houghton-Mifflin Co. Reprinted with permission.

EXERCISE 1: SEPARATION AND RATIO OF OIL AND ACID; EMULSIFIERS

PROCEDURE

1. Place vinegar and oil mixture in 3 test tubes as indicated.
2. Cover test tubes with plastic wrap and shake vigorously for 30 seconds.
3. Place test tube in rack. Record the time that elapses before the mixtures separate.

Vinegar	Oil	Time of Separation
2 t (10 ml)	2 t (10 ml)	
1 t (5 ml)	2 t (10 ml)	
1 t (5 ml)	3 t (15 ml)	

4. Repeat experiment with STANDARD and emulsifiers.

Emulsifier	Time of Separation
STANDARD: 2 t (10 ml) oil and 1 t (5 ml) vinegar	
STANDARD + ⅛ t (.63 ml) paprika	
STANDARD + ⅛ t (.63 ml) dry mustard	
STANDARD + ⅛ t (.63 ml) pepper	
STANDARD + Beaten egg yolk	

Conclusion:

EXERCISE 2: APPLICATION OF PRINCIPLES TO SALAD DRESSINGS

PROCEDURE

1. Prepare assigned salad dressings.
2. Serve dressings on salad greens.
3. Evaluate products as to consistency and flavor and general acceptability.

Dressing	Consistency	Flavor	General Acceptability
French			
Mayonnaise			
Cooked dressing			

SALAD DRESSING RECIPES

French Dressing

sugar	1 t	5 ml	vinegar	¼ c	60 ml
paprika	¼ t	1.25 ml	salad oil	½ c	120 ml
dry mustard	½ t	2.5 ml	clove garlic, crushed	1	1
pepper	¼ t	1.25 ml			

1. Place all ingredients in a jar. Cover, shake well, and refrigerate.
2. Shake immediately prior to serving. (¾ c; 180 ml)

Flavor variations:
    oil: safflower, canola, olive, sesame, hazelnut, walnut
    acid: cider, white or wine vinegar, balsamic, rice or fruit-flavored vinegars; citrus juice
    herbs/spices: parsley, celery seed, tarragon, horseradish, curry, Worcestershire sauce, etc.
    cheese: crumbed Roquefort, blue, parmesan

FRENCH DRESSING

Cooked Mayonnaise[1]

egg yolks	2	2	dry mustard	1 t	5 ml
vinegar	2 T	30 ml	pepper	dash	dash
water	2 T	30 ml	salad oil	1 c	240 ml
sugar	½ t	2.5 ml			

1. Place all ingredients except for oil in a double boiler over simmering water. Stir constantly until mixture bubbles in one or two places.
2. Remove from heat and stand 4 minutes.
3. Place cooked mixture in a blender and blend on high, or blend well with a whisk.
4. Very slowly, add oil and blend until mixture is thick and smooth.
5. Cover and refrigerate if not used immediately. Yield: 1¼ c (300 ml).

[1] Adapted from American Egg Board, Park Ridge, IL. 1991.

Cooked Dressing

dry mustard	½ t	2.5 ml	milk	¾ c	180 ml
sugar	1 T	15 ml	egg, slightly beaten	1	1
flour	2 T	30 ml	vinegar or lemon juice	¼ c	60 ml
paprika	⅛ t	.63 ml	margarine	1 T	15 ml

1. Mix dry ingredients, add milk and cook over direct heat, stirring until mixture boils.
2. Gradually add some of the hot starch to egg.
3. Add warmed egg–starch mixture to pan. Continue to cook over low heat until egg has thickened.
4. Gradually add vinegar and fat to mixture. Yield: 1 c (240 ml).

EXERCISE 3: FAT-FREE, FAT-REDUCED, AND FAT REPLACED PRODUCTS

A. Cost and Palatability of Fat-Free, Fat-Reduced, and Fat Replaced Products

PROCEDURE

Evaluate and record the palatability characteristics, caloric value and cost of the products assigned.

Product	kcal	Cost/svg	Palatability
Mayonnaise Fat-free			
Reduced-fat			
Salad dressing (mayo. type) Fat-free			
Reduced fat			
Salad dressing Fat-free			
Reduced-fat			
Frozen dessert Fat-free			
Reduced-fat			
fat-replacement			
Fat replacement + sugar substitute			
Cottage cheese Fat-free			
Reduced-fat			
Nonstick cooking spray			
Butter butter blend			
Margarine Reduced-fat			
Fat-free			

1. Discuss nutritional advantages of using fat-free and fat-reduced products in the diet.

2. Discuss the various ways in which consumers could incorporate fat-free and/or fat-reduced foods in their diet if given the following situations:
   a. Consumers purchasing foods at the grocery

   b. Consumers choosing foods at restaurants

   c. Consumers preparing foods from "scratch" at home

3. Study the labels of commercial reduced or fat free dressings and compare with regular products. What additional ingredients and additives are used in the reduced and/or no fat products?

B. Fat Replacement Labeling

PROCEDURE

1. Read labels on products (from Part A) containing fat replacements.
2. Identify fat replacements by name.

Product	Fat Substitute

Suggest other reduced fat products consumers might want to have available in the marketplace. Explain why replacements may not be possible in all the suggested products.

EXERCISE 4: COMPARISON OF DIETARY FATS

PROCEDURE

1. Complete chart, noting the percent saturated, unsaturated and monounsaturated fat.
2. Record observations regarding those fats that were used in laboratory food preparation (e.g., composition, flavor).

COMPARISON OF DIETARY FATS

Fat/Oil	% Saturated	% Polyunsaturated	% Monounsaturated
Canola oil			
Corn oil			
Olive oil			
Peanut oil			
Safflower oil			
Soybean oil			
Sunflower oil			
Butter			
Lard			
Margarine			
Vegetable shortening			

Observations:

SUMMARY QUESTIONS—FATS AND OILS

1. Define emulsion.

2. Define an emulsifier and how it functions.

3. When oil and vinegar are shaken together, which liquid is the *continuous phase*? *Dispersed phase*?

4. Distinguish between a temporary and permanent emulsion in an oil–vinegar mixture.

5. Identify food ingredients that have the potential to function as emulsifiers. List several processed foods which contain additives that function as emulsifiers (see Appendix E).

6. What constituent of eggs is the emulsifier?

7. What causes the emulsion in mayonnaise to break? How may the emulsion be reformed?

8. Distinguish among French dressing, mayonnaise and cooked dressing as to proportion of ingredients and the emulsifier used in preparation.

9. Other than salad dressing, which products prepared in class, contain ingredients that emulsify the fat?

10. Explain why dressings for tossed salad should be added and mixed with the vegetables *just* before serving the salad.

11. Describe hydrogenation, noting the changes that occur in degree of saturation of the fats hydrogenated.

12. Which dietary fat is lowest in saturated fat? Which is highest?

13. Explain the role of fats in the human diet.

14. Concerning fat consumption in the United States: what is the average percentage of total calories coming from fat? What is the recommended percent?

15. Briefly list several ways Americans could lower their fat intake.

16. Identify several fat replacements and possible limitations with their use in foods.

J. Sugars, Sweeteners

OBJECTIVES

To describe conditions prerequisite to crystallization of sugar solutions

To know and describe factors affecting the rate of crystallization and size of crystals in sugar products

To describe the relationship between boiling temperature, sugar concentration, and structure of sugar products

To describe and relate the effect of interfering agents to the structure of sugar products

To summarize key principles essential to obtaining a desirable sugar product

To demonstrate an understanding of and ability to apply key principles in the preparation of a sugar product

To evaluate the nutritive value of sugar products

To evaluate the uses and nutritive value of sugar substitutes

REFERENCES

ASSIGNED READINGS

TERMS

Solute	Crystallization	Carmelization	Inversion
Solvent	Crystalline	Viscosity	Hydrolysis
Solution	Seeding	Negative heat of solution	Invert sugar
Saturated	Nuclei	Heat of crystallization	Sugar substitute
Supersaturated	Amorphous	Interfering agent	

EXERCISE 1: METHODS OF INITIATING CRYSTALLIZATION[1]

PROCEDURE

Carry out or observe a demonstration of crystallization. (requires overnight refrigeration)

1. Prepare a highly *concentrated* solution of sodium (Na) thiosulfate by adding 260 g sodium thiosulfate crystals to 100 ml boiling water. Stir to dissolve.
2. Pour into three similar size beakers or other glass containers. Mark samples, cover, and refrigerate, undisturbed overnight.
3. Carefully remove containers from refrigerator (the solutions are now supersaturated).
4. Initiate crystallization in the manner indicated below in the chart.

Method of Initiating Crystallization	Observations on Speed of Crystallization and Size of Crystal
Add 1 crystal of Na thiosulfate. Leave beaker undisturbed.	
Add 1 crystal of Na thiosulfate. Shake beaker vigorously.	
No addition of Na thiosulfate. Shake beaker vigorously.	

1. Why was the solution cooled before crystallization was initiated?

2. What type of solution is necessary for crystallization to occur? Why?

3. Define the term seeding.

4. Does seeding affect crystal size? Why?

5. Does agitation affect crystal size? Why?

---

[1] Adapted from Halliday EG, Noble IT, *Food Chemistry and Cooking*. Chicago: University of Chicago Press, 1943.

EXERCISE 2: THE RELATIONSHIP OF SUGAR CONCENTRATION TO BOILING POINT

PROCEDURE

1. Calibrate thermometer.
2. Prepare proportions of sugar and water as directed.
3. Heat the mixtures in saucepans until boiling. Record the initial boiling point for each sample.
4. Continue boiling each solution to a temperature of 11°F (6°C) *above the boiling point of water.*
5. Immediately remove the solutions from heat. Cool slightly.
6. Using a glass measuring cup, measure the volume of each sugar solution. Record volumes. Reserve solutions for Exercise 3.

Water	Sugar	Initial Boiling Temperature	Final Boiling Temperature	Final Volume
A. 1 c (240 ml)	½ c (120 ml)			
B. ½ c (120 ml)	½ c (120 ml)			

CAUTION—SUGAR SYRUPS BURN

1. How do the initial volumes of the sugar solutions compare?

2. How do the initial boiling points compare? Why?

3. How do the final volumes of the solutions compare? Why?

4. Why does solution A take longer to reach the specified final boiling point?

5. Based on this experiment, are the final concentrations of the sample the same or different? Explain.

EXERCISE 3: EFFECT OF TEMPERATURE AND AGITATION ON CRYSTAL SIZE

PROCEDURE

1. In a large saucepan mix 2 c (480 ml) sugar and 1 c (240 ml) hot water with the hot sugar syrups from Exercise 2.
2. Bring the mixture to a boil. Cover for a few minutes. Remove the cover and continue boiling without stirring.
3. When the solution reaches 236°F (113.5°C), remove saucepan from heat and divide the solution into approximately three parts as follows:
    a. Pour ⅓ over a thermometer placed on a marble slab. Cool undisturbed to 110°F (43.5°C). Manipulate with broad spatula until crystallization occurs. Knead until soft then shape into patties.
    b. Pour ⅓ into another saucepan and immediately begin beating with a wooden spoon. Beat until crystallization occurs.
    c. Continue heating remaining ⅓ to 300°F (149°C). Immediately pour onto foil, making small wafer shapes.

Temperature When Agitated	Texture (e.g., grainy/smooth)	Appearance (e.g., color/shininess)
A. 110°F (43.5°C)		
B. Beaten immediately		
C. 300°F (149°C)		

1. Why were the crystal structures in A and B different? Describe the structures and account for the differences.

2. What was the final structure of C? Why did it differ from A and B?

3. Define "heat of crystallization." Was this observed in the experiments? If so, when?

EXERCISE 4: EFFECT OF INTERFERING AGENTS ON SUGAR STRUCTURE

PROCEDURE

1. Calibrate thermometer.
2. Prepare assigned recipes for crystalline or amorphous product. Circle interfering agents.
3. Display and evaluate all products. Note differences in structure.

PALATABILITY STANDARD

AMORPHOUS

Caramels: smooth texture
        no graininess
        chewy, not sticky
Brittle: smooth, hard
Butterscotch: hard, clear amber color

CRYSTALLINE

glossy
smooth, creamy texture
holds shape, yet soft

CANDY RECIPES

INGREDIENTS			METHOD
Old-Fashioned Butterscotch			1. Combine first three ingredients. Bring to a boil.
brown sugar	1 c	240 ml	2. Cover pan for a few minutes. Uncover.
light corn syrup	2 T	30 ml	3. Cook, stirring as little as possible, to 288°F (142°C).
water	½ c	120 ml	4. Remove from heat. Stir in fat.
margarine/butter	2 T	30 ml	5. Drop by teaspoonfuls on greased foil.
Fondant			1. Combine first three ingredients in a saucepan. Bring to a boil.
sugar	1 c	240 ml	2. Cover pan for a few minutes.
water	½ c	120 ml	3. Boil to 236°F (113.5°C) in uncovered pan.
cream of tartar	⅛ t	.63 ml	4. Pour on marble slab.
flavoring	⅛ t	.63 ml	5. Add flavoring, cool undisturbed to 110°F (43.5°C).
			6. Beat until firm.
Chocolate Fudge			1. Combine first four ingredients in a saucepan. Bring to a boil.
sugar	1 c	240 ml	2. Cover pan for a few minutes.
milk	⅜ c	75 ml	3. Boil to 234°F (112°C) in uncovered pan.
corn syrup	1 t	5 ml	4. Remove from heat. Pour onto marble slab. Add fat and vanilla. Do not mix.
chocolate	1 oz	28 g	5. Cool, undisturbed to 110°F (43.5°C).
margarine/butter	1 T	15 ml	6. Beat until thick and creamy.
vanilla	½ t	2.5 ml	7. Pour into greased 6-inch (15-cm) pan.

CANDY RECIPES

INGREDIENTS

Peanut Brittle

sugar	1 c	240 ml
light corn syrup	½ c	120 ml
water	2 T	30 ml
margarine/butter	1½ T	22.5 ml
peanuts	½ c	120 ml
vanilla	½ t	2.5 ml
baking soda	1 t	5 ml

METHOD

1. Combine first three ingredients. Bring to a boil.
2. Cover pan for a few minutes. Uncover
3. Heat to 238°F (114°C) stirring as little as possible.
4. Add fat and peanuts. Stir constantly until 295°F (147°C).
5. Add vanilla and soda. Stir.
6. Pour onto lightly greased foil, spreading syrup thin, but minimally.

Vanilla Caramels

sugar	½ c	120 ml
brown sugar	¼ c	60 ml
light corn syrup	¼ c	60 ml
milk	½ c	120 ml
evaporated milk	¼ c	60 ml
margarine/butter	2 T	30 ml
vanilla	½ t	2.5 ml

1. Combine all sugars and milk. Bring to a boil.
2. Cover pan for a few minutes. Uncover.
3. Boil, stirring with a wooden spoon, to 240°F (116°C).
4. Add evaporated milk and continue boiling until 248°F (120°C).
5. Remove from heat. Stir in fat and vanilla.
6. Pour into greased 6-inch (15-cm) pan.

---

1. Compare the color and texture of Fondant with that of the crystalline product Exercise 3, A. Account for the differences.

2. Compare the following interfering agents, noting how they foster small crystals in crystalline products or prevent crystallization in amorphous products.

Margarine/butter	
Milk	
Corn syrup	

3. How does the addition of baking soda contribute to the palatability of peanut brittle?

EXERCISE 5: SUGAR SUBSTITUTES, HIGH-INTENSITY SWEETENERS

PROCEDURE

1. Study labels of several noncaloric and other sugar substitutes and record ingredients below.
2. Evaluate the palatability of common sugar substitutes, comparing initial and after-taste with sweetness of standard sugar solution. Record observations in table.

	Label Ingred.	Sweetener/ 1 c (240 ml) water	Palatability		
			Kcal	Initial Taste	Aftertaste
Standard		2 t sugar			
Acesulfame K		1 pkt "Sweet-One"®			
Aspartame		1 pkt "Equal"®			
Fructose		2 pkt			
Saccharin		1 pkt "Sweet 'N Low"®			

1. Based on readings, list the current regulations that govern the use of these sugar substitutes in food products.

2. Identify additional sugar alcohols and noncaloric or high-intensity sweeteners that are used in food products, noting specific products.

SUMMARY QUESTIONS—SUGARS, SWEETENERS

1. Why is a burn from boiling sugar syrup more severe than a burn from boiling water? Explain.

2. List important factors, common to all crystalline candy recipes that influence the formation of small crystals.

3. As an uncovered sugar solution boils, why does the observed boiling point continue to rise? How does viscosity change?

4. In a fudge recipe, if the sugar is increased but the amount of liquids remains the same, how will the cooking time be affected?

5. In fudge preparation, if the end boiling point has been exceeded by 39°F (4°C), how can the product be corrected? Why is this possible?

6. In the preparation of peanut brittle, if the temperature exceeds 300°F (150°C) and a brownish black mixture develops, can the product be corrected? Explain.

7. Identify several chemical interfering agents.

8. This graph illustrates the effect of interfering agents on the speed of crystallization of a sucrose solution.

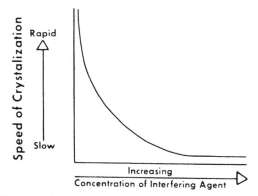

Source: Division of Nutritional Sciences, New York State College of Human Ecology at Cornell.

   a. What is the effect of an interfering agent on the speed of crystallization?

   b. Why is a somewhat slower rate of crystallization helpful in keeping crystals small?

9. In candy preparation, fudge fails to harden, and the texture is unsatisfactory. How might a new brand of sugar, 51% sucrose and 49% fructose, be the cause of this product failure? Would the results have been unsatisfactory if this sugar had been used in the preparation of peanut brittle?

10. Applying the principles of crystallization to the preparation of ice cream (freezing), predict how the following variables would affect crystal size.

Variable	Crystal Size
Slow rate of freezing	
No agitation	
Low freezing temperature	
Substitution of cream for milk	

11. List various substitutes and the brand names under which they are sold:
    a. Sugar alcohols

    b. Noncaloric or high-intensity sweeteners

12. Discuss why sugar substitutes may not be effective in all prepared foods as a major ingredient. What precautions must be observed?

13. Identify advantages and disadvantages of using sugar substitutes, considering cost, function and nutritive value.

14. Identify sugar substitute, reduced-sugar or sugar-free products that also use fat substitute, reduced-fat, or fat-free formulations. Identify advantages and disadvantages of using these products.

Part III

Heating Foods by Microwave

Microwave heating and cooking have been added to the traditional ways foods are processed and cooked. Microwaves, similar to radio and TV energy waves, are absorbed by a food, especially the water, fat and sugar components. The molecules in the food vibrate at a high rate of speed, billions of times per second, and heat results from this friction. The heat is thus produced instantaneously in the food and not by transfer from an outside heat source, as occurs with other heat transfer methods.

Microwave cooking complements conventional cooking methods, but does not replace them for cooking all foods. Microwave cooking may keep the food preparation area cooler, be quicker and more convenient for our modern lifestyles. Microwave ready dinners, entrees or side dishes are readily available in grocery stores. Major nutrients, as well as important palatability characteristics, are well preserved in microwave cooking.

When one learns how to use a microwave oven, including use of the probe, memory capacity, and various power level options correctly, it can be a valuable cooking aid.

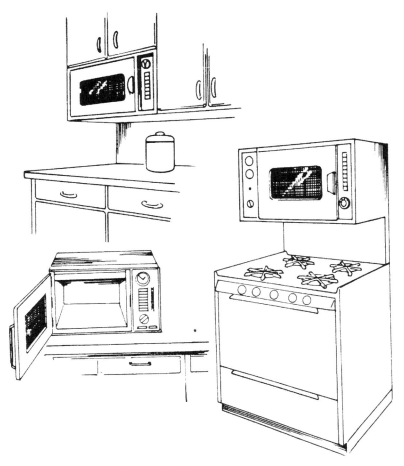

Source: Olson W, Olson R. North Central Regional Publ. 70-1979. Agric. Extension Service, University of Minnesota

Microwave Cooking

OBJECTIVES

To describe how microwave energy heats food
To utilize microwave cooking procedures for heating a variety of foods
To compare microwave cooking procedures, utensils, and palatability of final products to conventional cooking
To follow a variety of package instructions for microwavable products
To delineate the advantages and disadvantages of microwave cooking

REFERENCES

Appendices G-I, G-II; H

ASSIGNED READINGS

TERMS

Molecular Friction	Thermal Runaway	Hot Spots	Magnetron
Ionizing/Nonionizing	Heat Penetration	Shielding	

The following exercises using the microwave oven may be completed individually or incorporated into the various exercises of Part II in the manual. Microwave on HIGH setting unless otherwise noted.

Doubling a recipe quantity may require 1½ times as long; *halving* a recipe requires approximately ⅔ time.

PRICK FOODS TO RELEASE PRESSURE

Courtesy: General Electric Company

ARRANGE FOODS IN A CIRCLE. IF PORTIONS ARE IRREGULARLY SHAPED OR SIZED, PLACE THICKEST PORTIONS TO THE OUTSIDE OF THE DISH

Courtesy: General Electric Company

EXERCISE 1: EFFECT OF COOKING PROCEDURE ON PIGMENTS AND FLAVORS

Procedure

1. Prepare 1 c (240 ml) of assigned vegetables representing pigments and flavor compounds; place in small glass container.
2. Add ¼ c (60 ml) water and cover tightly with plastic wrap.
3. Microwave for 3 minutes. Remove ½ c (120 ml) vegetable.
4. Cover tightly and microwave 5 minutes longer.
5. Drain, label and display both samples. Record observations.

Pigments	Cooked 3 minutes		Cooked 8 minutes	
	Color	Texture	Color	Texture
Chlorophyll				
Anthocyanin				
Flavor	Flavor	Texture	Flavor	Texture
Allium				
Brassica				

Conclusions:

EXERCISE 2: FRUITS

PROCEDURE

1. Prepare fruits as directed.
2. Label and display. Record observations.

Applesauce

apples, pared, cored, sliced	2	2	sugar	1 T	15 ml
water	2 T	30 ml	nutmeg, cinnamon	dash	dash

1. Place apples and water in small casserole, cover.
2. Microwave 4 minutes, or until tender.
3. Stand, covered for 2 minutes. Add sugar and spices, mash, if necessary.

Baked Apples

apples, cored	2	2	water	2 T	30 ml

1. Slit skin around top center of fruit to prevent bursting during cooking. Place apples in small casserole.
2. Add water, cover and microwave for 6 minutes or until tender.
3. Stand, covered for 2 minutes.

Fruit	Palatability
Applesauce	
Baked apple	

EXERCISE 3: VEGETABLES

GENERAL DIRECTIONS

1. Prepare vegetables by cutting small, uniform pieces, pricking skin or arranging with more tender parts toward the center of the dish.
2. Add approximately ¼ c (60 ml) water per pound (454 g) of vegetable.
3. Cover and microwave all vegetables, turning or stirring vegetable halfway through cooking.
4. Evaluate palatability of finished product.

PROCEDURE

1. Follow recipe directions for assigned product.

2. Evaluate palatability and, if possible, compare to like products cooked by conventional methods.

Vegetable	Palatability
Baked potato	
Pennsylvania red cabbage	
Cauliflower	
Baked eggplant with tomato sauce	
Italian vegetable medley	
Baked tomato	
Packaged frozen vegetable; use microwave package instructions	

Baked Potatoes

1. Prick scrubbed potato in two or three places and place on a paper towel in oven.
2. Microwave 5 to 7 minutes, turning potato once or twice during cooking.
3. Let stand, uncovered 2 minutes before cutting open. Open and top with seasonings.

Pennsylvania Red Cabbage

1. Use ingredients listed in the Fruits and Vegetables chapter.
2. In a 2-quart (2-L) baking dish combine all ingredients except vinegar.
3. Cover. Cook 6 minutes. Stir and microwave 6 minutes longer. Test cabbage for desired degree of doneness; stir in vinegar. If necessary, microwave longer. Stand 2 to 3 minutes. (2 to 3 servings)

Cauliflower

cauliflower flowerets	2 c	480 ml	margarine		1 t	5 ml
water	2 t	10 ml	dill weed		season	season

1. Place cauliflower in 1 qt (1 L) baking dish.
2. Add water, cover and cook 6 minutes. Turn dish and continue microwaving 2 to 3 minutes until cauliflower tests done. Stand 2 to 3 minutes.
3. Add seasonings desired. (2 servings)

Baked Eggplant With Tomato Sauce

medium eggplant	½	½	oregano	½ t	2.5 ml
tomato sauce	1 c	240 ml	mozzarella cheese slices	2 1-oz	57 g

1. Pare eggplant; cut into 1/8-inch (.3-cm) thick slices.
2. Spread 2 T (30 ml) sauce on bottom of 1-quart (1-L) casserole.
3. Mix remaining sauce and oregano. Layer eggplant and sauce in casserole.
4. Cover and cook 4 minutes. Rotate dish ¼ turn and cook 4 minutes longer. Test to see whether vegetable is tender. If not, cook 2 minutes longer.
5. Place mozzarella cheese on top of eggplant and cook 1 minute longer, until cheese has melted. Stand 2 to 3 minutes. (3 to 4 servings)

Italian Vegetable Medley

broccoli flowerets	1½ c	360 ml	sliced carrot rounds	½ c	120 ml
cauliflower flowerets	1½ c	360 ml	green peppers	2	2
fresh mushrooms	4 oz	114 g	(1 inch [2.54 cm] squares)		
zucchini, sliced thin	1 c	240 ml	Italian dressing	¼ c	60 ml

1. Arrange vegetables in a 2-quart (2-L) glass serving dish or pie plate; place broccoli flower side up, around outside; then cauliflower, zucchini, peppers, mushrooms, and finally carrots in center.
2. Spoon Italian dressing evenly over all vegetables.
3. Cover and cook 3 minutes or until vegetables are tender. (6 to 8 servings)

Baked Tomatoes

1. Follow recipe in Fruits and Vegetables chapter for ingredients.
2. In glass cup, place fat and onion, cover and cook 1 minute. Mix with bread crumbs and seasonings. Stuff tomatoes.
3. Place stuffed tomato halves in a 1 qt (1 L) glass casserole dish. Cover.
4. Cook, giving dish half-turn after half of cooking time. Two halves take 2 to 2½ minutes; 4 halves, 3 to 4 minutes.

EXERCISE 4: STARCH PRODUCTS

PROCEDURE

1. Cook starch items as assigned.
2. Label and display. Record observations.

Starch Product	Palatability
Oatmeal	
Farina	
Basic sauce	
Milk-based sauce	
Cinnamon sugar sauce	
Pudding Butterscotch	
Chocolate	
Packaged product; use microwave package instructions	

A. Pasta, Rice, and Cereals

Pasta and rice take about the same time to cook on the stove and in the microwave oven. Microwave reheating however, is an excellent way to reheat pasta or rice. To reheat, cover dish tightly with lid and microwave. (Open lid carefully.)

Oatmeal

oatmeal   ⅓ c   80 ml   water   ¾ c   180 ml

Place ingredients in a 1-quart (1-L) glass container; cover and cook 3 to 4 minutes.

Cream of Wheat or Farina

| cereal | 2½ T | 37.5 ml | water | 1 c | 240 ml |

Place ingredients in a 1-quart (1-L) glass container; cover and cook 3 to 4 minutes.

B. FLOUR AND CORNSTARCH AS THICKENERS

Basic Starch Thickened Sauce

| margarine | 2 T | 30 ml | milk | 1 c | 240 ml |
| flour | 2 T | 30 ml | | | |

1. Place fat and flour in a 1 qt (1 L) casserole. Cook for 2 minutes, stirring after 1 minute.
2. Gradually add milk, stirring.
3. Cook 3½ to 4 minutes, stirring every minute until mixture boils. Yield: 1 c (240 ml).

Variations for Milk-Based Sauce

Cheese Sauce: To finished sauce add 1 to 2 oz (28 to 57 g) grated cheddar cheese and a dash of cayenne pepper. Cook 1 minute to melt cheese.

Curry Sauce: Add 1 t (5 ml) curry powder with flour and proceed as directed in Basic Sauce.

Cinnamon Sugar Sauce

sugar	½ c	120 ml	water	1 c	240 ml
cornstarch	1½ T	22.5 ml	margarine	1 t	5 ml
cinnamon	1 t	1.25 ml			

1. Mix ½ of sugar, cornstarch, cinnamon and water in a small glass container.
2. Cover and cook 3 to 4 minutes until mixture boils, stirring sauce after 1½ minutes.
3. Add remaining sugar and fat, stir until blended. Yield: 1⅓ c (320 ml).

Variation:
   Citrus Sauce: omit cinnamon, add 1 T (15 ml) lemon juice and 1 t (5 ml) lemon rind in step 3.

Butterscotch and Chocolate Puddings

Use *ingredients* listed in Cereal and Starch chapter, to make puddings. Control addition of ingredients that could adversely affect thickness of starch sol. Follow *directions* in Basic Starch Thickened Sauce recipe (above).

MICROWAVE COOKING

EXERCISE 5: EGGS[1]

CAUTION:	Never hard cook eggs or reheat eggs that are in the shell. Pressure builds up inside and eggs burst.

PROCEDURE

1. Prepare egg products as directed.
2. Label and display. Record observations.

Egg Product	Palatability
Scrambled	
Baked custard	
Fried—Browning Dish	

Scrambled Eggs

| margarine | 2 t | 10 ml | eggs | 2 | 2 |
| milk | 2 T | 30 ml | | | |

1. Place margarine in glass bowl or 2-c (480-ml) measuring cup; microwave on HIGH until melted (about 30 seconds).
2. Add milk, mix. Add eggs, beat with a fork.
3. Cook for 45 seconds on HIGH; stir set portions from outside to center.
4. Cook 45 seconds on MEDIUM; repeat stirring. When finished, eggs should be firm and be 165°F (74°C). Stand 1 to 2 minutes.

Baked Custard

| milk | 1 c | 240 ml | sugar | 1 T | 15 ml |
| egg | 1 | 1 | vanilla | ½ t | 2.5 ml |

1. Beat ingredients until well mixed.
2. Pour mixture into three 6 oz (180 ml) glass cups.
3. Cook 3 minutes on MEDIUM; rotate oven position and cook for 3 more minutes on MEDIUM. Test for doneness by inserting a clean knife. Continuing cooking and testing at 30-second intervals until egg is coagulated. Custard should not boil.

Eggs on Browning Dish

Following manufacturer's directions, fry an egg on a browning dish. Check temperature of cooked product to ensure safety.

---

[1] Check final temperature of product. DO NOT taste any egg product that does not reach a final cooking temperature of 140°F (60°C), held for 3½ minutes, or 160°F (71°C).

EXERCISE 6: MEAT, POULTRY, AND FISH[1]

PROCEDURE

1. Prepare products as directed.
2. Microwave fish on HIGH, meat and poultry on MED-HIGH for best results.
3. Label and display. Record observations.

Product	Palatability
Meatballs	
Chicken	
Baked fish	
Packaged product use microwave package instructions	

Baked Fish

fish fillet	½ lb	227 g	margarine	2 t	10 ml

1. Arrange fish in glass baking dish; dot with margarine.
2. Cover dish with double thickness of wet paper towels.
3. Microwave 4 minutes, rotate dish ½ turn, cook 3 more minutes. Test to see if fish flakes. Continue cooking for 1-minute intervals until fish flakes. (2–3 servings)
   Toppings: sweet-sour, tomato, barbecue sauce.

Chicken

margarine	2 t	10 ml	paprika, herbs	season	season
chicken thighs	2	2			

1. Place margarine in microwave-safe container; microwave until melted.
2. Place chicken thighs in container, turning to coat with margarine. Arrange meatiest parts toward outside of dish.
3. Sprinkle with seasonings and cover with wax paper.
4. Microwave 5 to 7 minutes on MEDIUM-HIGH. Let stand, covered 5 minutes.
   (1 piece: 3 to 5 minutes; 2½ to 3 lb [1.14 to 1.36 kg]: 22 to 29 minutes).
   Chicken is done when clear juice runs from a fork prick.

---

[1] Review information in Appendices G-I and G-II concerning regulations about cooking temperatures for meat, and poultry.

Meatballs

ground beef	½ lb	227 g	bread crumbs	¼ c	60 ml
egg	1	1	finely chopped onion	2 T	30 ml

1. Mix ingredients. Shape into 6 balls.
2. Arrange balls in a circle on a glass pie pan. Cover.
3. Microwave 5 to 8 minutes rotating dish half-turn after 3 minutes. Microwave 4 minutes, test for doneness. Continue cooking for 30-second intervals until finished. Stand, covered 2 to 3 minutes. (2 servings)

EXERCISE 7: BATTERS AND DOUGHS

PROCEDURE

1. Prepare baked products as directed at various power levels.
2. Baked products may be prepared with herbs, spices, or other ingredients in order to provide color.
3. Label and display. Record observations.

Baked Product	Palatability
Muffins	
Biscuits	
Pastry shell	
Packaged product use microwave package instructions	

Conclusions:

Muffins

1. Use ingredients in BASIC MUFFIN recipe, Batters and Doughs chapter.
2. Once mixed, pour batter into six slightly greased 6 oz (180 ml) glass custard cups or microwave muffin pan, filling cups ½ full.
3. Arrange cups in a ring or on a microwave muffin pan.
4. Microwave 3 to 5 minutes on MEDIUM-HIGH. Check for doneness at 2½ minutes. Continue to bake at 30 second intervals until done. Rotate ½ turn at each 30 second interval. Stand 2 minutes. (Muffins will seem barely set and top may have moist spots, but toothpick inserted in center comes out clean when done.) (6 muffins)

Biscuits

1. Use ingredients in BASIC BISCUIT recipe, Batters and Doughs chapter.
2. Cut dough into 12 biscuits.
3. Place a drinking glass or glass cup in center of greased glass pie plate.
4. Arrange biscuits around drinking glass, overlapping to fit.
5. Microwave 6 to 8 minutes on MEDIUM, rotating dish half-turn after 3 minutes. Stand 2 minutes. (12 biscuits)

Basic Pastry Shell

| sifted flour | 1 c | 240 ml | water | | 2 T | 30 ml |
| shortening | ⅓ c | 80 ml | herbs, spices | | to color | |

1. Mix ingredients by pastry method. Roll and fit into glass pie plate; prick pastry with fork.
2. Microwave 6 to 7 minutes rotating half-turn after 3 minutes. Check for doneness: Botton will be dry and opaque; top dry, blistered, but not brown.

EXERCISE 8: REHEATING BAKED PRODUCTS

PROCEDURE

1. Place 5 baked yeast rolls in a circle on a paper towel in the microwave oven.
2. Microwave 10 seconds remove 1 roll. Repeat at 4 second intervals, taking out a roll at each interval.
3. Compare the overall palatability of the reheated rolls.
4. Summarize results of experiment.

Reheated Rolls	Results

EXERCISE 9: DEFROSTING

PROCEDURE

1. Place two 1/4-lb (114-g) ground beef patties on separate plates.
2. Microwave one patty on HIGH for 4 minutes. Microwave the other patty for 4 minutes on DEFROST (or LOW).
3. Observe the internal quality of the meat. Evaluate the extent of defrosting or cooking that has occurred.
4. Summarize results below.

Defrosting Method	Quality
High	
Defrost	

SUMMARY QUESTIONS—MICROWAVE COOKING

1. Why do microwave recipes often require that the container be turned during the cooking process?

2. Why is a "rest period" or "standing time" included in directions for microwave recipes?

3. List several ways that the lack of browning in microwave cooking may be overcome.

4. From readings, compare nutrient retention in microwaved vegetables to vegetables cooked by conventional methods.

5. Considering sanitary quality, why do the USDA directions recommend to microwave pork to 170°F (77°C)?

6. In defrosting foods, why is a low setting used and not high?

7. Do containers used for microwave cooking get hot? Explain.

8. Based on experiments, summarize the palatability characteristics of the following microwaved products:

Fruits	
Vegetables	
Starches	
Eggs	
Meats, poultry, fish	
Batters, doughs	

9. How do directions on packaged microwaveable food products assist the consumer? Note any suggestions for improvements in package directions.

10. State some recipe changes that must be made in converting cooking from a conventional oven to a microwave oven.

11. Are all foods successfully prepared in the microwave oven? Specify what quality standards are not met using microwave cooking?

12. Provide an example of when an *ingredient* for a recipe, but not the *whole* recipe, may be cooked in the microwave oven.

PART IV

Meal Management

What makes a nutritious meal an enjoyable one? Generally, we want a meal to be personally satisfying and obtainable within our resources of time, money, knowledge, and energy.

The consideration of how economics, individual taste, lifestyle, cultural and ethnic background, and special nutritional needs, affects meal planning for individuals and groups.

The suggested meal management projects are an opportunity for you to apply creatively the principles of food selection and preparation.

Courtesy: R.T. French Co.

Meal Management

OBJECTIVES

To apply principles of nutrition, sanitary quality, economics, and the science of food to meal planning
To adapt meal plans to a variety of cultures, including international and regional domestic menu patterns
To adapt meal plans to low cost, low calorie and other modifications
To demonstrate basic food preparation skills, use of equipment, time management, and service of food

REFERENCES

Appendices A, B, C, D, G-I, G-II, L, M, N

ASSIGNED READINGS

EXERCISE 1: ANALYZING MENUS FOR PALATABILITY QUALITIES

PROCEDURE

Using the following chart of menus, identify planning errors and then make suggestions for improvements in the palatability of the meals. Where relevant, also note how nutritional aspects could be improved.

Menu	Planning Errors	Improvements
Ground beef sauté Stewed tomatoes Lima beans Parslied buttered rice Applesauce Tea		
BBQ pork on a bun French fries Buttered carrots Broccoli spears with cheese sauce Baked rice pudding Milk		
Curried eggs on rice Lettuce wedge, vinaigrette dressing Roll and butter Gingerbread Nonfat milk		
Ham-lentil soup, crackers Grilled cream cheese sandwich Waldorf salad Apple turnover with cheese Milk		
Cream of chicken soup, crackers Cottage cheese and sliced peach salad Baked custard Milk		

EXERCISE 2: ECONOMIC CONSIDERATIONS IN MENU PLANNING

PROCEDURE

Adapt the following high-cost foods to moderate- and low-income budgets, as possible.

	Moderate-Cost Food	Low-Cost Food
Calves' liver		
Ground sirloin		
Fresh orange juice		
Frozen halibut		
Fresh salmon		
Center slice ham		
Porterhouse steak		
Packaged baked sweet rolls		
Fresh tomatoes		
Whole milk, fluid		
Aged sharp cheddar cheese		
Fresh asparagus tips		
Leaf lettuce		
Butter		
Frosted corn flakes		
Frozen chocolate cream pie		
Packaged hash brown potatoes		
Frozen pancakes		

SUGGESTIONS FOR MEAL PLANNING PROBLEM

Low calorie
Low cholesterol
Low sodium
High complex carbohydrate
Low income

Moderate income
Regional United States
International
Fat—30% of calories
Oven meals

Stovetop meals
Vegetarian
Lacto-ovo-vegetarian
Preschool, nursing home

EXERCISE 3: LOW-CALORIE MODIFICATIONS

PROCEDURE

The following menu[a] totals approximately 2,600 to 2,700 calories. Adapt this plan for an individual who wishes to consume 1600 to 1700 calories.

		Modifications
Breakfast		
Orange juice (fresh)	¾ c (180 ml)	
Scrambled egg	1 lg	
Bacon	1 sl	
Bagel with cream cheese	1 2 T (30 ml)	
Jam	1 t (5 ml)	
Milk, whole	1 c (240 ml)	
Water, tea or coffee		
Brown bag lunch		
Ham sandwich, sliced ham	3 oz (85 g)	
lettuce	2 leaves	
mayonnaise	1 T (15 ml)	
whole wheat bread	4 slices	
Bean salad; French dressing	1 c (240 ml) 2 t (10 ml)	
Chocolate chip cookie	1	
Apple	1 medium	
Blueberry yogurt	1 c (240 ml)	
Water, tea or coffee		
Dinner		
Vegetable chowder, milk base	1 c (240 ml)	
Baked fish with tomato sauce	5 oz (140 g) ⅓ c (80 ml)	
Buttered broccoli spears	½ c (120 ml)	
Mixed green salad: lettuce, spinach, green onions, cucumbers	1½ c (360 ml)	
Blue cheese dressing	2 T (30 ml)	
Gingerbread	1 svg	
Pear (fresh)	1 med	

[a]Adapted from *Ideas for Better Eating*, USDA, 1981.

EXERCISE 4: MEAL PLANNING

PROCEDURE

1. Within the framework of the defined problem, plan a FULL DAY'S MENU. (LUNCH or DINNER will be prepared in laboratory.)

   a. Special Meal Planning Problem:

   b. Nutritive Value:
      1. Plan a full day's menu that meets your RDA. Check tentative plan using the criteria of the Food Guide Pyramid or other appropriate food guide. Complete nutritive calculations.
      2. Calculate calories, protein, calcium, iron, ascorbic acid, vitamin A, fat, cholesterol and sodium. (Percent of calories from carbohydrate, protein, and fat may be calculated.)
         NOTE: Single foods fortified to levels that meet 100% of RDA may not be used, although such products may be included, but only 5 mg iron/serving counted.

   c. Cost: day/person   lunch/person
                     breakfast/person   dinner/person

   d. Palatability:
      Menu should be planned to take into account food preferences, color, texture, shapes, flavor, and temperature of foods.

   e. Time: The meal (lunch or dinner) is to be prepared in (   ) hours.

2. Prepare a report including the following: Report Due Date:
   a. Market Order and special equipment for meal you plan to prepare (Worksheet A).
   b. Planning Schedule (Worksheet B).
      Menu to be prepared.
      Work-time schedule.
      Serving plan.
   c. Nutritive Value of Day's Menu (Worksheet C).
   d. Copies of recipes for meal you are preparing.

3. LUNCH or DINNER will be prepared on:

4. Analysis of Criteria Used (Worksheet D) Due Date:
   Exercise 5.

WORKSHEET A
MARKET AND EQUIPMENT ORDER

Materials Needed	Amount Needed	Cost
Vegetables		
Fruits		
Meats		
Eggs		
Milk and milk products		
Cereal products		
Fats		
Sugars		
Miscellaneous		
Equipment		

Worksheet B
Planning Schedule

Name:

Date of Meal Preparation:

Menu of Meal Prepared:

1. Work out a time schedule for the preparation of your meal.

Time	Preparation Steps

2. Diagram complete table setting (cover), noting placement of food on plates and serving dishes, serving utensils, and centerpiece.

3. Indicate procedure for serving and clearing cover, if you plan two or more courses.

WORKSHEET C
NUTRITIVE VALUE OF DAY'S MENU[a,b]
(CIRCLE MEAL PREPARED: LUNCH–DINNER)

Food	Measure	Energy (kcal)	Protein (g)	Calcium (mg)	Iron (mg)	Vitamin C	Vitamin A	Fat (g)	Cholesterol (mg)	Sodium (mg)

[a] Indicate subtotals for each meal.
[b] A nutrient analysis software program may be used.

Worksheet C (Cont'd)

Food	Measure	Energy (kcal)	Protein (g)	Calcium (mg)	Iron (mg)	Vitamin C	Vitamin A	Fat (g)	Cholesterol (mg)	Sodium (mg)
Total for Day										

Percentage Kcal from:

Protein

Carbohydrate

Fat

MEAL MANAGEMENT 243

WORKSHEET D
SUMMARY ANALYSIS OF MEAL PLAN

Evaluate your completed meal plan, noting the criteria you established and the various strategies you used to meet the criteria. Where appropriate include specific examples.

Criteria	Strategies

Use additional pages if necessary for a complete analysis.

EXERCISE 5: MEAL PREPARATION

PROCEDURE

1. Prepare meal with assigned classmate(s) to meet approved time schedule.
2. Serve meal.
3. Evaluate the completed meal, noting specific examples of strategies employed to achieve a successful meal.
4. Instructor will evaluate the following:
   - Adherence to time schedule
   - Food science principles
   - Cooking techniques
   - Palatability of meal
   - Table setting and ease of service
   - Kitchen organization and clean up
   - Use of equipment
   - Completeness of self-evaluation

STUDENT EVALUATION

MENU FOR DAY

Breakfast   Lunch   Dinner

1. Complete Budget of Time and Money:

Factor	Budgeted	Expended	Comments
Time			
Money			

2. Evaluate palatability:

Characteristic	Good	Fair	Poor	Suggestions for Improvement
Color				
Texture				
Shape or form				
Temperature				
Satiety				
Flavor				
Degree of doneness				

3. List sanitary precautions taken during the preparation and service of the meal.

4. Select one cooked product prepared for the meal and analyze the recipe for the application of food science principles:

Product Name:

Step	Principles

SUMMARY QUESTIONS—MEAL MANAGEMENT

1. Why are the aesthetics of menu planning and service important?

2. In planning, which nutrients must be planned in specific amounts?

3. The RDA for iron is frequently difficult to meet. How can the absorption of iron be enhanced? What dietary factors inhibit iron absorption?

4. Identify advantages and disadvantages of using established food guides (e.g., Food Guide Pyramid, Dietary Guidelines) for meal planning.

5. In planning meals for low income levels, which foods or food groups could be increased (because of their excellent nutritive value for dollars spent)? Which could be decreased?

6. Identify factors, other than cost, which may make planning low-income meals difficult.

7. Provide examples of how the inclusion of fast foods in the diet may impact an individual's food choices in menu planning.

8. What are some of the physical limitations that make preparing and/or eating common foods difficult for some individuals?

9. What are the nutritional criteria for a satisfactory breakfast? Devise a breakfast which meets this criteria for a dieter, a vegetarian and a person who never has enough time to eat breakfast.

10. Make a checklist to evaluate sanitation practices to be used when you prepare food. Consider food buying, storage and preparation of meal preparation.

11. Indicate on thermometer the temperature or range of temperatures for the following:

   a. lukewarm, scalding, simmering, poaching

   b. gelatinization of wheat/corn starch

   c. coagulation of whole egg

   d. end boiling temperature of sugar syrup for crystalline product

   e. end boiling temperature of sugar syrup for amorphous product

   f. oven temperature for baking: soufflé in waterbath; soufflé without waterbath; biscuits; popovers; pastry

   g. internal temperature for rare, medium, and well-done meat

   h. temperature for holding of foods

   Source: Division of Nutritional Sciences, New York State College of Human Ecology at Cornell.

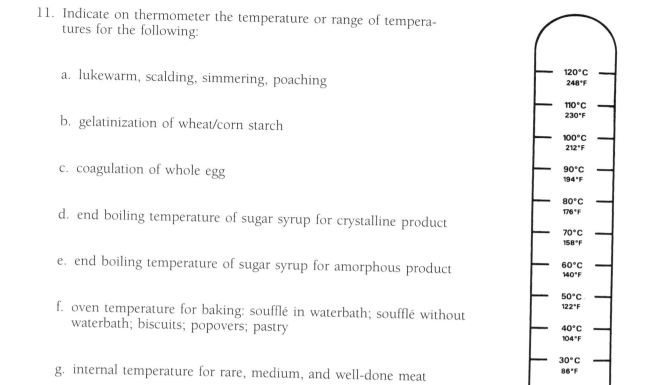

Appendix A
Legislation Governing the Food Supply

LAWS ENFORCED BY THE FOOD AND DRUG ADMINISTRATION[1]

The Food and Drug Administration is a federal regulatory agency responsible for enforcing laws to protect consumers of foods, drugs, cosmetics, medical devices, chemical products, and other articles used in the home. Congress enacts the laws and relies on the FDA to establish necessary regulations and standards, and, thereafter to enforce the regulations and standards.

THE FEDERAL FOOD, DRUG AND COSMETIC ACT—1938

The Food, Drug, and Cosmetic Act replaced the original Food and Drug Act of 1906 with new and stronger provisions. The following regulations of this law refer to foods;

- Food must be pure and wholesome, safe to eat, and produced under sanitary conditions. Adulterations and misbranding are defined.
- Labeling must be truthful and informative
- Provisions are stated for establishing standards of identity, quality, and fill.

Standards of Identity prevent adulteration by defining exactly what a specific food must contain. For example, fruit jams must contain at least 45 parts of fruit and 55 parts of sugar or other sweetener.

Standards of Quality set minimum specifications for such factors as tenderness, color, and freedom from defects in canned fruits and vegetables. For example, quality standards for canned foods limit the "string" in green bean, excessive peel in tomatoes, hardness in peas, and "soupiness" in cream-style corn.

Standards of Quality should not be confused with Grades A, B, C, Prime, Choice, Fancy, and so forth, which are set by the USDA. Manufacturers pay USDA for this voluntary service.

Standards of Fill tell the packer how full the container must be to avoid deception of the consumer and charges of "slack filling".

The law provides for enforcement through inspection, collection of samples and prosecution in the courts.

MILLER PESTICIDE AMENDMENT—1954

This amendment sets safety limits for pesticide residues allowed on raw agricultural commodities.

[1] Condensed and adapted from FDA.

Poultry Products Inspection Act—1957

This Act regulates poultry and poultry products, including labeling.

Food Additive Amendment—1958

Key points of this amendment are:

- Prohibits the use of new additives until manufacturer has established its safety
- Establishes a list of additives, generally recognized as safe (GRAS). This list was drawn up after a review of available evidence showed no significant risk from the intended use of the additive.

Since 1970, the FDA has been reassessing the safety of additives on the GRAS list. A panel of scientists, established under an FDA contract with the Federation of American Societies for Experimental Biology (FASEB) has reviewed all experimental evidence to determine if continued use of additives on the GRAS list is justified. Consumers can expect the safety assessment of additives to continue and perhaps the GRAS list will be revised in light of additional scientific data.

- Contains a statement known as the Delaney Clause. This clause states that no food additive can be approved by the FDA for human food if it is found to induce cancer when ingested by man or animal. The Delaney Clause does not recognize any level of cancer-producing chemical as safe.

Color Additives Amendment—1960

This amendment allows FDA to establish by regulation the condition for safe use of color additives in food.

Fair Packaging and Labeling Act—1966

This act requires that consumer products in interstate commerce be honestly and informatively labeled. FDA is empowered to enforce provisions which affect foods, drugs, cosmetics, and medical devices. The Federal Trade Commission (FTC) administer the law with regard to other products.

Because there are existing federal laws governing the labeling of meat products, poultry, and poultry products, these foods are excluded from the provisions of the Fair Packaging and Labeling Act and are monitored by the USDA.

Some specific points of this act in regard to food labels are:

- The label must state the name and address of manufacturer, packer, or distributor of the food.
- The label must show the common or usual name of the product.
- The ingredients must be listed on the label in descending order of their predominance by weight in the product. Exceptions are some foods for which standards of identity have been established. The ingredient descriptions of standardized foods are on file with the FDA.
- Additives used in food products must be listed on labels. This ruling also applies to standardized foods, with the exception of butter, cheese, and ice cream.
- A statement of the net weight or volume of contents must appear on the main display label of the package.
- Weights between 1 pound and 4 pounds net weight must also be stated in total number of ounces, so cost per ounce can be figured easily.
- Labels must not carry misleading terms that qualify units of measure such as "giant quart" or "jumbo pound."
- FDA has authority to limit the amount of packaging material or air space to the amount that is required to protect the contents of the package, or which is required by the kind of machinery used to package the commodity.

FOOD LABELING REGULATIONS—1973

Label regulations were established by the FDA to provide the consumer with valuable tools for identifying and selecting nutritious foods. The components of the program were interrelated and called for several new concepts in food labeling, such as identifying and giving amounts of nutrients in a food product, establishing nutritional quality guidelines for certain foods or giving the percentage of the characteristic component in a product.

The regulation specified methods and formats of labeling products intended *for special dietary needs.* It established a standard of identity for *dietary supplements,* including vitamins, minerals and highly enriched foods. This regulation required use of "imitation" when a food was nutritionally inferior to a food product for which it is a substitute. Foods containing any amount of artificial flavor were to be labeled with the name of the food and characterizing flavor preceded by the words "artificial" or "artificially flavored."

NUTRITION LABELING AND EDUCATION ACT (NLEA)—1990

As a result of the NLEA there are regulations which specify information food processors must include on their labels, including Nutrition Facts. There is extensive, mandatory nutrition labeling of food, standard serving sizes, and use of health claims. The purpose of the NLEA is to:

- Assist consumers in selecting foods that can lead to a healthier diet
- Eliminate consumer confusion
- Encourage production innovation by the food industry

The FDA set 139 reference serving sizes for use on Nutrition Facts labels that more closely approximates amounts consumers actually eat that previous labeling. General descriptive terms allowed for use on food labels were provided.

1992: The FDA's voluntary point-of-purchase nutrition information for raw produce and fish

1994: The Food Safety and Inspection Service (FSIS) of the USDA introduced regulation for voluntary nutrition labeling of raw meat and poultry. May 8, 1994 was the deadline for meeting NLEA requirements for labeling with "Nutrition Facts" panel on packaging.

PATHOGEN REDUCTION: HAZARD ANALYSIS AND CRITICAL CONTROL POINT (HACCP) SYSTEM REGULATION—1996

This regulation codifies principles for the prevention and reduction of pathogens and requires the development of Sanitation Standard Operating Procedures (SSOP's) and a written HACCP plan that is monitored and verified by inspectors of meat and poultry processing plants.

Appendix B
Food Guides and Dietary Guidelines

FOOD GUIDES

Food grouping systems are designed to provide a simple, organized method of helping people plan an adequate and balanced diet. The most widely used system today is the Food Guide Pyramid introduced in 1992 by the United States Department of Agriculture (USDA). In this guide, foods are classified into one of five groups depending on the similarity of their nutritive value. Each group provides some but not all of the nutrients needed to maintain good health, and no one group is more important than another. The Food Guide Pyramid lists the groups and the recommended number of servings from each group. Serving sizes within each group provide approximately similar amounts of the major nutrients. The tip of the Pyramid contains Fats, Oils and Sweets which is *not* a major food group.

USDA FOOD GUIDE PYRAMID

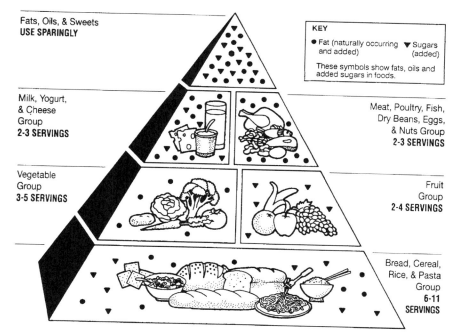

Source: U. S. Department of Agriculture

Because most foods can readily be placed in one of the food groups the Food Guide Pyramid easily adapts to usual eating patterns, as well as various cultural and ethnic food choices. For example, tortillas and grits would be classified under the Bread, Cereal, Rice, and Pasta Group; and papayas and mango under Fruit. Some mixed foods, e.g., macaroni and cheese, need to be broken down to their component parts for placement in food groups; macaroni would be placed in the Bread, Cereal, Rice, and Pasta Group and cheese in the Milk, Cheese, and Yogurt Group.

OTHER FOOD GUIDE PYRAMIDS

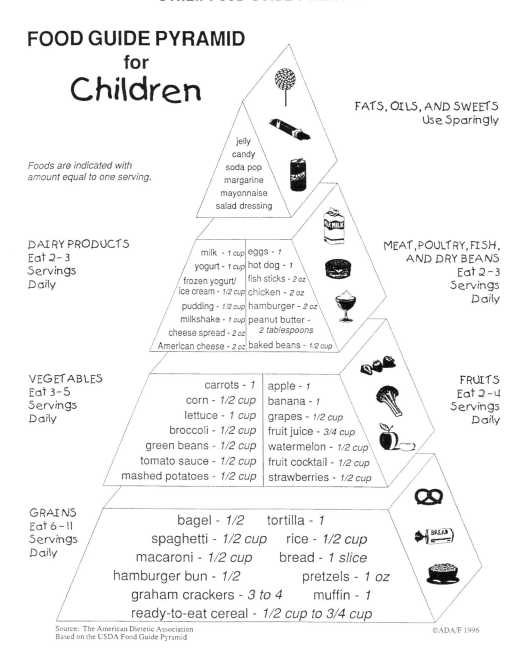

Source: The American Dietetic Association
Based on the USDA Food Guide Pyramid
©ADA/F 1996

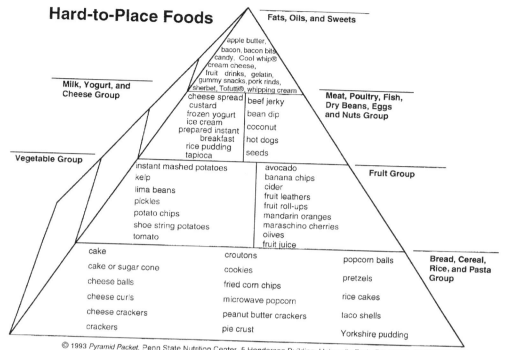

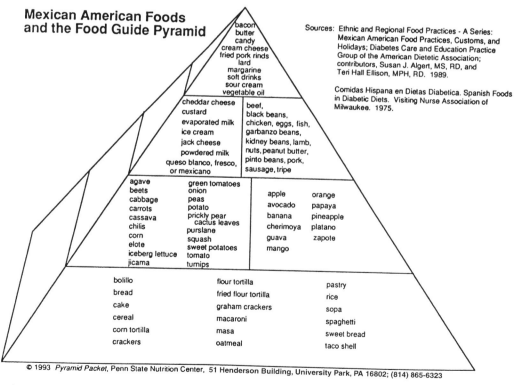

Additional multicultural food guides are available from Penn State Nutritional Center, including African-American, Chinese, Indian, Jewish, Navajo, Puerto Rican and Vietnamese Foods.

DIETARY GUIDELINES

Many Americans are concerned that their diets are too high in energy, fat, sugar and sodium, and too low in fiber. These dietary imbalances have been associated with high incidence of chronic diseases such as heart disease and cancer. In response to this concern, the USDA and the Department of Health and Human Services developed the Dietary Guidelines for Americans. The guidelines are designed for healthy people over two years old.

These guidelines stress good eating habits based on variety and moderation. In contrast to the Food Guide Pyramid which suggests food choices to obtain an adequate supply of all nutrients, the Guidelines were developed to help moderate dietary excesses. Used together, the Food Guide Pyramid and Dietary Guidelines are simple, and effective tools to help the selection of a well-balanced diet without excess.

Dietary Guidelines for Americans

- Eat a variety of foods.
- Balance the food you eat with physical activity; maintain or improve your weight.
- Choose a diet with plenty of grain products, vegetables, and fruits.
- Choose a diet low in fat, saturated fat, and cholesterol.
- Choose a diet moderate in sugars.
- Choose a diet moderate in salt and sodium.
- If you drink alcoholic beverages, do so in moderation.

Appendix C
Some Food Equivalents

Protein Equivalents[1] (16–20 g)		
milk	2¼ c	540 ml
cottage cheese	½ c	120 ml
cheddar type cheese	2½ oz	70 g
meat	3 oz	85 g
eggs	3	3
cooked dried beans/peas	1 c	240 ml
peanut butter	5 T	75 g
hot dogs	2½	2½
walnuts, almonds, cashews	¾ c	180 ml
bread slices	8	8
prepared dry cereal	8 oz	227 g
cooked macaroni	3 c	720 ml
cooked rice	4 c	950 ml

Vitamin C Equivalents (60–70 mg)		
orange	1	1
orange juice	½ c	120 ml
grapefruit	¾ c	180 ml
cantaloupe	½	½
strawberries, whole	¾ c	180 ml
watermelon	1/16	1/16
green pepper	¾	¾
tomato	1½	1½
tomato juice	1½ c	360 ml
kale, turnip greens, cooked	1 c	240 ml
collards, cooked	¾ c	180 ml
broccoli, cooked	½ c	120 ml
spinach, cabbage, cooked	1¼ c	300 ml
potatoes, white or sweet	3	3

Calcium Equivalents[2] (270–320 mg)		
whole milk	1 c	240 ml
nonfat milk	1 c	240 ml
buttermilk	1 c	240 ml
yogurt	1 c	240 ml
ice cream	1½ c	360 ml
cheddar cheese	1⅓ oz	37 g
creamed cottage cheese	1¼ c	300 ml
cream cheese	1 lb	454 g
oranges	6	6
ground beef	6 lb	2.72 kg
bread slices, enriched	14	14
cooked broccoli, kale	2 c	480 ml
cooked collard greens	1 c	240 ml

Iron Equivalents[3] (4–5 mg)		
whole grain enriched bread	8 slices	
cooked dried beans/peas	1 c	240 ml
ground beef	4 oz	114 g
liver	1½ oz	43 g
clams, canned	4 oz	114 g
eggs	4	4
plums, canned	2 c	480 ml
prune juice	½ c	120 ml
dried apricots	½ c	120 ml
tomato juice	2 c	480 ml
peas, canned	1 c	240 ml
peas, cooked	1½ c	360 ml
spinach, cooked	1 c	240 ml
white potatoes, medium	5	5

[1] Protein equivalent refers to *amount* of protein, not quality of protein.
[2] Spinach and rhubarb have substantial amounts of calcium, but it is complexed with oxalate and not fully utilized.
[3] Availability of iron in eggs and plant sources may be less than indicated.

Appendix D
Average Serving or Portion of Foods[1,2]

Item	Weight/Amount		Item	Weight/Amount	
A. Dairy Products			**F. Fruits**		
Cheese			Fresh	By the piece	
Cheddar	1–2 oz	28–57 g	Fresh, cut up: canned	½ c	120 ml
Cottage	¼–½ c	60–120 ml	Fruit juices	½ c	120 ml
Milk			**G. Meat**		
To drink	1 c	240 ml	Fresh, frozen, canned, ckd	2–3 oz	57–85 g
For cooked cereal	2 oz	60 ml	Liver	2 oz	57 g
For dry cereal	4 oz	120 ml	Bacon	2 slices	
Ice cream	½ c	120 ml	Frankfurters	2	2
Yogurt	½–1 c	120–240 ml	Lunch meat	2 slices	
B. Eggs			**H. Plant Protein**		
Fried, poached, hard or soft cooked	1	1	Legumes, beans, cooked	1½ c	360 ml
Scrambled	1½	1½	Peanut butter	2–4 T	30–60 ml
			Nuts, seeds	¼ c	60 ml
C. Fats and Oils			**I. Poultry, cooked**		
Butter or Margarine	2 t	10 ml	Chicken, turkey, boned	3 oz	85 g
Salad Dressing	2 T	30 ml	Chicken, broiler	½ bird	
Mayonnaise	1 T	15 ml	Chicken, fryer	¼ bird	
D. Fish, Shellfish			**J. Vegetables**		
Cooked	2–3 oz	57–85 g	Cooked, canned	½ c	120 ml
E. Bread and Cereals			Lettuce	¼ head	
Bread, rolls, muffins	1	1	Asparagus	3–6 spears	
Cereals, cooked	½–¾ c	120–180 ml	Brussels sprouts	4–6	4–6
Cereals, dry	¾–1 c	28 g	Corn, ears	1	1
Macaroni, rice, cooked	½ c	120 ml	Potatoes, whole (medium)	1	1
Saltines	4	4	Potatoes, french fries	8–10	8–10
Graham crackers	4	4			
Pancakes and waffles	2–3	2–3			

[1]These may not be amounts personally consumed. Knowledge of the average quantity per serving is useful in estimating cost and nutritive value. See references for more information.
[2]See FDA standard serving sizes on food labels for individual foods.

Appendix E
Food Additives

An additive is a "a substance or a mixture of substances, other than a basic foodstuff, which is present in a food as a result of an aspect of production, processing, storage or packaging" (National Research Council Food Protection Committee).

Every food additive used in processing should serve one or more of the following purposes:

- improve or maintain nutritional value
- enhance quality
- reduce wastage
- enhance consumer acceptability
- improve keeping quality
- make the food more readily available
- facilitate preparation of the food

In its broadest sense, a food additive is any substance added to food. By legal definition it is "any substance the intended use which results or may reasonably be expected to result—directly or indirectly—in its becoming a component or otherwise affecting the characteristics of any food". Two *exempt* categories from the food additive regulation process include GRAS and prior sanctioned substances are *not legally* considered food additives.

Many people think of any additive added to foods as complex chemical compounds, although salt, baking soda, and vanilla are commonly used in foods today. The common *lay usage* of the term "food additive" differs from the *legal definition*.

FUNCTIONS OF FOOD ADDITIVES

Basic functions of some food additives are described below. For each classification, examples of additives and their uses in specific foods are provided.

Antioxidants—Antioxidants combine with available oxygen and are added to halt oxidation reactions, to prevent rancidity in fats, oils, cereals, crackers, potato chips and other foods, and to extend shelf life. They prevent or inhibit oxidation of unsaturated fats and oils, colors and flavorings. Ascorbic acid and the tocopherols are naturally occurring antioxidants. Synthetic antioxidants include butylated hydroxyanisole (BHA), butylated hydroxytoluene (BHT), tertiarybutylhydroxyquinone (TBHQ) and propyl gallate.

Nitrites function as antioxidants to fix the color, flavor and stability of cured meat.

Bleaching and Maturing Agents—Freshly milled flour has a yellowish color and relatively poor baking quality. If stored for several months, the flour "ages," that is, whitens in color and improves in baking quality. Because natural aging is slow and costly and insect or rodent infestation difficult to control during storage, chemicals are added to speed up the process. Benzoyl peroxide exerts only a bleaching action. The oxides of nitrogen, chlorine dioxide, nitrosyl chloride and chlorine both bleach and mature flour.

Bulking Agents—Bulking agents such as sorbitol, glycerol, and polydextrose (glucose, sorbitol and citric acid in 89:10:1 ratio) are used in small amounts to provide body, smoothness and creaminess which supplement the viscosity and thickening properties of hydrocolloids.

Coloring Agents—Food colors are used to make processed food look more appetizing by imparting a characteristic color. Baked goods, candy, dairy products (e.g., butter, margarine, and ice cream), gelatin desserts, and jams and jellies often contain color additives. Natural food colors used include annato extract (yellow), cranberries, beet juice, tomatoes (red), carotene, and saffron (yellow-orange), and tumeric.

Curing Agents—Curing agents impart color and flavor to foods such as bacon, frankfurters, ham and salami. They also have antimicrobial properties which lowers the temperature needed to kill *Clostridium botulinum*. They inhibit the growth of *Clostridium perfringens, Staphylococcus aureus,* and nonpathogens.

Edible Films and Waxes—Edible films such as the polysaccharides cellulose, pectin, starch, and vegetable gums, or proteins such as casein and gelatin may be applied with a thin coat, to foods. Edible waxes are applied to fruits and vegetables to improve or maintain appearance, prevent mold, and contain moisture, while still allowing respiration. Food grade *vegetable waxes*, including petroleum-, beeswax-, and shellac-based wax or resin, and food grade *animal-based* waxes are regulated as GRAS.

Emulsifiers—Emulsifiers are surface-active agents and have the ability to surround small droplets of fat, thereby dispersing them throughout a mixture. Emulsifiers are used in cake mixes, confections, ice cream, salad dressings and shortenings to improve uniformity of performance.

Some common emulsifiers are mono- and di-glycerides, lecithin and polysorbates. The presence of emulsifiers affects texture of starch products and they are sometimes included to improve the texture of dehydrated potatoes and to help retain the softness of bread.

Enzymes—Enzymes are non-toxic substances that occur naturally in foods, catalyzing various reactions. They are easily inactivated by pH and temperature. Bromelain (from pineapple), ficin (from figs), and papain (from papaya) are used as meat tenderizers. Amylases hydrolyze starch in flour, invertase is used to hydrolyze sucrose in candy, pectinases clarify pectin-containing jellies or juices, and proteases are used in cheesemaking and soy sauce production.

Fat Replacers—Fat replacers include carbohydrate-, fat-, and protein-based substances such as maltodextrins, sucrose polyesters of fatty acids and sucrose, and microparticulated protein, respectively.

Firming Agents—Firming agents such as calcium chloride improve processed fruits or vegetables by hardening or firming the texture.

Flavoring Agents—Flavoring agent, both natural and synthetic, make up the largest group of food additives and are used in baked products, confections, ice cream, prepared meats, and soft drinks. Natural flavoring substances include herbs, spices (e.g., salt, pepper, cloves, ginger) and sweeteners, essential plant oils (citrus), and extractives (vanilla extract). Synthetic flavors include amyl acetate (banana), benzaldehyde (almond, cherry), citral (lemon), and flavor enhancers such as monosodium glutamate (MSG).

Humectants—Humectants or moisturizing agents prevent such foods as coconut and candy from drying. Examples include polyhydric alcohols such as glycerol, propylene glycol, mannitol, and sorbitol that are used to improve texture and retain moisture because of their affinity for water.

Nutrient Supplements—Historically the term *enrichment* has denoted the process of adding vitamins and minerals to processed foods to compensate for losses incurred during processing, storage, and distribution.

Fortification has referred to the addition of a nutrient deemed lacking in the diet, to an appropriate food. "Except for foods with specific standards, the two terms often are used interchangeably in food labels".[1] The FDA defines both terms as the addition of nutrients to food.[2]

Thiamin, riboflavin, niacin, iron and folate[3] and in some instances calcium and vitamin D, are added to milled grains. The FDA has established standards of maximum and minimum amounts of these nutrients for enrichment of corn grits, cornmeal, farina, macaroni, noodle products, rice, and wheat flour.

Vitamins A and D may be added to margarine; Vitamin D to both fluid and evaporated milk; Vitamins A and D to fluid and nonfat dry milk. In many regions of the United States, iodine is added to table salt.

pH Control Substances—Natural or synthetic acid or alkali ingredients change or maintain the initial pH of a product. For example, the tart taste of soft and fruit drinks is achieved through the use of organic acids, natural or synthetic. Acids, alkalis, buffers or neutralizing agents may also be used as flavor additives or to preserve food. For example, acid ingredients lower the pH of foods and inhibit microbial growth. Acetic acid, citric acid or organic acids from apples or figs (malic acid), lactic acid or tartaric acid may be useful as additives, the latter for leavening. Calcium propionate is added to control the pH of breads.

Preservatives—Food additives may be used to delay natural deterioration, not to disguise it. Food can deteriorate through microbial growth of molds, bacteria and yeast, and through reaction with oxygen which may alter flavor, color and texture. Inhibitors such as vinegar, salt, and sugar are used in pickles, sauerkraut, jams, and jellies. The vinegar is acidic, and the salt and sugar compete with bacteria for water and therefore lower the water activity (A_w). Other additives with these functions are calcium or sodium propionates, and potassium sorbate are additives used to control mold in bread and bacilli growth which causes "rope" in breads or mold. Sodium benzoate inhibits yeast and mold in confections, fruit juices, margarine, and pickles.

A preservative may be used alone, or in combination with other additives or preservation techniques such as cold or heat preservation, or dehydration.

Sequestrants—Sequestrants such as EDTA (ethylenediaminetetracetic acid) and pyrophosphate form inactive complexes with metallic ions that can catalyze fat oxidation or the formation of cloudy precipitates. Additionally, they prevent metals from catalyzing reactions of pigment discoloration, flavor or odor loss, or vitamin oxidation. Sequestrants are added to carbonated beverages, cooked hams, margarine, salad dressings, canned shrimp and tuna, and vinegar.

Stabilizers and Thickeners—Stabilizers and thickeners are used to give a smooth, uniform texture to many foods. The presence of stabilizers and thickeners prevents the separation of chocolate particles in chocolate milk, keeps ice crystals smaller in ice cream, imparts "body" to artificially sweetened beverages and maintains uniform texture in puddings and confections.

[1] Thonney P, Bisogni C. *Fortified and Enriched Foods. You Should Know About Food Ingredients*, 6, No. 2. Division of Nutritional Sciences. Ithaca, NY: Cornell University, 1983.
[2] *Federal Register,* January 25, 1980.
[3] *Federal Register,* March 5, 1996.

Included in this group of additives are alginates (from kelp), carrageenan (a seaweed derivative), dextrins of starch and modified starches; hydrocolloids (material that holds water) such as gelatin (e.g., the protein from animal bones, hoofs), vegetable gums such as gum tragacanth, gum arabic, guar, and locust bean, and pectin; and cellulose compounds such as methylcellulose, carboxymethylcellulose (CMC), and sorbitol.

Other Miscellaneous Additives

Anticaking agents, dough conditioners, fumigants, leavening agents, lubricants, propellants, and *artificial* and *natural sweeteners* are also regarded as food additives.

Appendix F
pH of Some Common Foods[1]

2.0	Limes	5.2	Turnips, cabbage, squash
2.1		5.3	Parsnips, beets
2.2	Lemons	5.4	Sweet potatoes, bread
2.3		5.5	Spinach
2.4		5.6	Asparagus, cauliflower
2.5		5.7	
2.6		5.8	Meat, ripened
2.7		5.9	
2.8		6.0	Tuna
2.9	Vinegar, plums	6.1	Potatoes
3.0	Gooseberries	6.2	Peas
3.1	Prunes, apples, grapefruit (3.0–3.3)	6.3	Corn, oysters, dates
3.2	Rhubarb, dill pickles	6.4	Egg yolk
3.3	Apricots, blackberries	6.5	
3.4	Strawberries, lowest acidity for jelly	6.6	Milk (6.5–6.7)
3.5	Peaches	6.7	
3.6	Raspberries, sauerkraut	6.8	
3.7	Blueberries, oranges (3.1–4.1)	6.9	Shrimp
3.8	Sweet cherries	7.0	Meat, unripened
3.9	Pears	7.1	
4.0	Acid fondant, acidophilus milk	7.2	
4.1	Commercial mayonnaise (3.0–4.1)	7.3	
4.2	Tomatoes (4.0–4.6)	7.4	
4.3		7.5	
4.4	Lowest acidity for processing at 100°C	7.6	
4.5	Buttermilk	7.7	
4.6	Bananas, egg albumin, figs, isoelectric	7.8	
4.7	point for casein, pimientos	7.9	
4.8		8.0	Egg white (7.0–9.0)
4.9		8.1	
5.0	Pumpkins, carrots	8.2	
5.1	Cucumbers	8.3	

[1]Reprinted by permission from *Handbook of Food Preparation*, 6th Ed. American Home Economics Association, Washington, DC. 1975.

Appendix G-I
Major Bacterial Foodborne Illnesses[a]

	Salmonellosis (Infection)	*Staphylococcus* (Intoxication)	*Perfringens* (Infection/Intoxication)	Botulism (Intoxication)
Causes	*Salmonella* (facultative) Bacteria widespread in nature, live and grow in intestinal tracts of human beings and animals	*Staphylococcus aureus* (facultative) Bacteria fairly resistant to heat; bacterial toxin produced in food is extremely resistant to heat. Toxin produces illness.	*Clostridium perfringens* (anaerobic) Spore-former. Vegetative cells destroyed with thorough cooking; spores can survive to germinate, and numbers grow.	*Clostridium botulinum* (anaerobic) Spore-forming organisms that grow and produce a potent neurotoxin.
Symptoms	Severe headache, followed by vomiting, diarrhea, abdominal cramps, and fever. Infants, elderly, and persons with low resistance most susceptible. May cause death in these groups.	Vomiting, diarrhea, prostration, abdominal cramps.	Nausea without vomiting, diarrhea, acute inflamation of stomach and intestines.	Dizziness, double vision, inability to swallow, speech difficulty, progressive respiratory paralysis. Fatality rate is high unless diagnosed promptly and an antitoxin given.
Onset	6–72 hours	1–6 hours	8–22 hours	12–36 hours
Duration	2–3 days	24–48 hours	24 hours or less	Fatal if untreated
Source	Transmitted by eating contaminated food, or by contact with infected persons or carriers of the infection. Also transmitted by insects, rodents, and pets; domestic and wild animals.	Transferred to foods by humans from hands, nasal passages, infection and skin abrasions.	Note: Caused by large numbers of this bacteria (infection) in a food which, after ingested, produces a toxin in the gut (intoxication).	Toxin in food.
Foods	Eggs; poultry, red meats, unpasteurized dairy products	Custards; eggs; meat and meat products; warmed-over foods	Large cuts of meats, stews, soups, or gravies that have been kept on steam tables for long periods of time or cooled slowly.	Canned low-acid foods; smoked fish; stews; honey for infants; baked potatoes in foil; large quantities of sauteed vegetables kept unrefrigerated overnight.
Prevention	Avoid cross-contamination. Cook all meats thoroughly, and cook poultry and eggs to 165°F (74°C). Cool quickly.	Proper heating and refrigeration. Good personal hygiene. Toxin is destroyed only by boiling several hours or in a pressure cooker for 30 minutes at 240°F (115.5°C)	Time-temperature control in cooling and reheating meat. Maintain foods out of the temperatures between 45°F (7°C) and 140°F (60°C).	Bacterial spores destroyed by high temperatures obtained only in a pressure canner. (More than 6 hours at boiling point needed to kill spores.) Toxins destroyed with 10–20 minutes of boiling, depending on food density. Keep sous-vide packages refrigerated.

[a]Cross-contamination spreads bacteria to other foods. Sanitize hands, work surfaces and utensils.

continued

Appendix G-I (cont'd)
Major Bacterial Foodborne Illnesses

	Listeriosis (Infection)	*Bacillus cereus* (Intoxication)	Campylobacteriosis (Infection)	*Escherichia coli* 0157:H7 (Infection/Intoxication)
Causes	*Listeria monocytogenes* Facultative; can grow in damp environment	*Bacillus cereus* Facultative.	*Campylobacter jejuni*	*Escherichia coli* 0157:H7
Symptoms	Meningitis in immuno-compromised individuals. Nausea, vomiting, headache, fever, chills.	Nausea, vomiting, diarrhea, abdominal cramps.	Diarrhea, fever, headache, nausea, abdominal pain.	Bloody diarrhea, diarrhea, nausea, severe abdominal pain, vomiting, occasionally fever.
Onset	1 day to 3 weeks	8–16 hours (can be ½–5 hours)	3–5 days	12–72 hours
Duration	Indefinite. High fatality in immuno-compromised individuals.	less than 12 hours	1–4 days	1–3 days
Source	Humans; also domestic wild animals, soil, water, mud	Soil and dust	Domestic and wild animals	Humans (intestinal tract), cattle and other animals
Foods	Unpasteurized milk and some soft cheese products; raw meat/poultry; chilled, ready-to-eat foods; raw vegetables	Cooked rice and rice dishes, and cereal products; food mixtures; spices; sauces; vegetable dishes; meatloaf. Especially a problem where large batches of foods are prepared ahead and improperly cooked/reheated.	Unpasteurized milk and dairy products; untreated water; raw vegetables; undercooked meats.	Raw and undercooked red meats, unpasteurized milk, and fruit juices, cream-baked pies
Prevention	Use only pasteurized milk and dairy products. Cook foods to proper temperature.	Time and temperature control: quick chilling, reheating to 165°F (74°C).	Avoid unpasteurized milk and untreated water. Cook foods thoroughly. Wash hands thoroughly after handling raw poultry and meats.	Avoid cross-contamination. Cook ground beef to 155°F (68°C). Good personal hygiene.

Appendix G-II
Meat and Egg Cooking Regulations

FDA Model Food Code:

Poultry	165°F (74°C)	Pork	150°F (66°C)
Ground beef	155°F (68°C)	Rare roast beef	130°F (54°C)
Eggs	140°F (60°C), for 3½ minutes, or 160°F (71°C)	Reheating (except rare roast beef 130°F [54°C])	165°F (74°C)

Appendix H
Heat Transfer

Radiation. Energy waves travel through space and heat the surface of food (broiling, toasting).

Conduction. Heat is transferred from molecule to molecule (pan or burner).

Convection. Warmed air or water rises, creating currents which heat surface of food (air in oven, liquid in pan).

Microwave. Electromagnetic waves penetrate food and attract and repel molecules.

Most foods are heated by a combination of heat transfer methods. The following photographs show how heat is transferred in a potato. The lightest area indicates the hottest part of the potato.

HEATING FOODS BY MICROWAVE

Microwaves are high frequency electromagnetic waves and like radio broadcast waves, do not break chemical bonds (unlike ultraviolet light, X-rays and gamma rays). When produced by magnetron in a microwave oven, they penetrate the food ¾ to 1 inch (1.9 to 2.54 cm). The microwaves oscillate or alternate millions of times each second, and in doing so, attract and repel the polarized molecules in the food. The molecules in the food vibrate and create *friction,* which produces the *heat energy* that cooks the food.

A given amount of microwave energy is emitted by the magnetron, and this is divided among the foods placed in the oven. Doubling the amount of food placed in the oven *nearly* doubles the heating time. Areas of food not reached by the microwaves, as well as the containers, eventually are heated by *conduction* from the food heated by microwaves.

CAUTION!

DO NOT OPERATE OVEN WHEN EMPTY!!

DO NOT OPERATE OVEN

IF DOOR IS DAMAGED!!

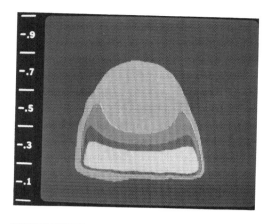

RANGE TOP. AFTER 8 MINUTES, HEAT HAS BEEN CONDUCTED FROM BOTTOM, BUT THE LARGE TOP AREA OF POTATO IS UNCOOKED.

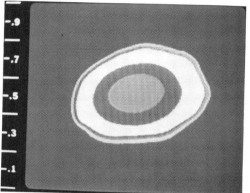

OVEN. AFTER 15 MINUTES, THE HEAT FROM SURFACE IS BEING CONDUCTED TO INTERIOR; THE CENTER IS UNCOOKED.

MICROWAVE. AFTER 4 MINUTES, THE POTATO IS HEATED THROUGHOUT BY MICROWAVE RADIATION AND CONDUCTION.

Courtesy: General Electric Co.

A. Factors Affecting Cooking Times (Olson and Olson 1979)
   1. Temperature of food—frozen foods should be defrosted on low settings, prior to cooking; refrigerated food requires more time than the same room temperature food.
   2. Density—denser foods require longer cooking
   3. Moisture—more moisture requires longer cooking (more heat):
       75–90% (most heat): vegetable soups, casserole
       50–60% (less heat): meat, poultry, fish
       20–35% (least heat): breads, cakes
   4. Sugar and Fat—more requires less cooking time

5. Shape—thin slices cook faster than thick, chunky foods; round shapes cook more evenly than square or rectangular shapes.

B. Utensils Used in Microwave Ovens.
1. MAY USE: Paper towels (caution with food-contact by second-hand, recycled paper), plastic wrap (although no criteria for "microwave safe" are established), wax paper, glass and glass-ceramic, pottery or china with no metallic contents trim or glaze
2. To test *if uncertain*: Place 1 c (240 ml) water in glass and put it on or in a dish to be tested. Microwave 1 minute on HIGH. If water becomes hot, dish is safe (not absorbing microwave energy), but if *dish* is hot, do not use.
3. AVOID: Metal pans, thermometers, skewers, foil trays, Corning Centura, Melamine dishes. Plastics vary in their ability to withstand microwave energy.

C. Browning.
1. Microwaving is moist heat cooking, therefore browning is difficult to achieve, unless a browning utensil, special browning unit or broiler is used in conjunction with microwaves.
2. Many microwave recipes add special toppings to improve surface appearance.
   Quick breads, yeast breads—toasted seeds, nuts or coconut, cinnamon sugar, herb seasonings
   Meats—soup mixes, sauces such as soy or barbeque, browning and seasoning; crushed chips, or seasoned crumbs, paprika; microwave preparations to shake on.
   Casseroles—toasted bread crumbs, cheese, sauces, fried dry onion rings, etc.

For consumer information on microwave oven radiation, consult Consumer Service Department of specific oven manufacturers or write: Microwave Ovens, HFX-28, Bureau of Radiological Health, Food and Drug Administration, Rockville, Maryland 20857; your State health department; or your local FDA office.

REFERENCE

Olson W, and Olson R. *Heating Prepared Foods in Microwave Ovens.* North Central Regional Extension Publ. No. 72, Agricultural Extension Service, University of Minnesota, 1979.

Appendix I
Symbols for Measurements and Weights

t or tsp = teaspoon	pk = peck	oz = ounce	g = gram
T or Tbsp = tablespoon	pt = pint	lb = pound	kg = kilogram
c = cup	qt = quart	fl = fluid	ml = milliliter
bu = bushel	gal = gallon	fd = few drops	l = liter

MEASUREMENT EQUIVALENTS

(for laboratory use)

1 T = 100 drops
    = 15 ml
    = 3 t

1 c = 16 T
    = 8 fl oz
    = 237 ml (240 ml)
    = ½ pt

1 qt = 4 c
    = 32 fl oz
    = 2 pt
    = 946.2 ml

1 gal = 4 qt
    = 3.8 l

1 kilo = 1000 g
    = 2.2 lb

1 lb = 16 oz
    = 453.6 g (454 g)
    = 0.45 kg

1 g = 0.035 oz
    = 1000 mg

1 oz = 28.35 g (28 g)
3½ oz = 100 g

TEMPERATURE CONVERSIONS

$°F = 9/5°C + 32$
$°C = 5/9 (°F − 32)$

Appendix J
Notes on Test for Presence of Ascorbic Acid

Most measurements of ascorbic acid are based on its oxidation-reduction properties. When a substance *loses* hydrogen atoms or electrons, it is *oxidized*. The substance *gaining* the electrons is reduced. Carbon atoms 2 and 3 of ascorbic acid easily lose their hydrogen atoms to appropriate substances, forming dehydroascorbic acid.

$$\text{Ascorbic acid (reduced form)} \underset{+2H}{\overset{-2H}{\rightleftarrows}} \text{Dehydroascorbic acid (oxidized form)}$$

For the test a specific dye, 2,6-dichlorophenolindo-phenol, is used. It is purplish blue in oxidized form, changing to light pink or becoming colorless when it is reduced. When solutions of ascorbic acid and the dye are reacted, the ascorbic acid gives up its hydrogen atoms to the dye. Ascorbic acid is oxidized and the dye is reduced. The reaction is indicated by change in color of the dye as shown below:

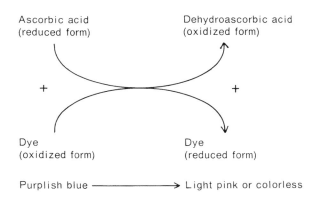

Other substances such as some sugars can also react with the dye. This test used with the foods suggested is a crude indicator of ascorbic acid concentration and the changes it undergoes during processing treatments.

Appendix K
Cooking Terms[1]

Bake. To cook in an oven or oven-type appliance. Covered or uncovered containers may be used.

Barbecue. To roast slowly on a spit or rack over coals or under a gas broiler flame or electric broiler unit, usually basting with a highly seasoned sauce. The term is commonly applied to foods cooked in or served with barbecue sauce.

Baste. To moisten food while cooking by pouring over it melted fat, drippings, or other liquid.

Blanch. To heat for a short period of time in boiling water or steam (precook).

Boil. To cook in water, or liquid sauce, at boiling temperature (212°F at sea level; 100°C). Bubbles rise continually and break on the surface.

Braise. To cook slowly in a moist atmosphere. The cooking is done in a tightly covered utensil with little or no added liquid. Meat may be browned in a small amount of fat before braising.

Broil. To cook uncovered by direct heat on a rack placed under the source of heat or over an open fire.

    Pan Broil. To cook in lightly greased or ungreased heavy pan on top of range. Fat is poured off as it accumulates, food does not fry.

Caramelize. To heat sugar or food containing sugar until a brown color and characteristic flavor develop.

Cream. To work a food or a combination of foods until soft and creamy, using a spoon, paddle, or other implement. Most often applied to fat or a mixture containing fat, for example, shortening and sugar.

Cut in. To distribute solid fat into dry ingredients using two knives or a pastry blender.

Fold. To combine two mixtures, or two ingredients such as beaten egg white and sugar, by cutting down gently through one side of the mixture with a spatula or other implement, bringing the spatula along the bottom of the mixture, and then folding over. This motion is repeated until the mixture is well blended.

Fricassee. To braise individual pieces of meat, poultry, or game in a little liquid—water, broth, or sauce.

[1] Adapted by permission from: *Family Fare,* Home and Garden Bulletin No. 1, USDA, 1973.

Fry. To cook in fat without water or cover.

    Pan-fry or sauté. To cook in a small amount of fat (a few T [or 45 ml] or more, up to ½ inch (1.25 cm)) in a fry pan.

    Deep-fry or french-fry. To cook in a deep kettle, in enough fat to cover or float food.

Grill. Same as broil.

Knead. To press, stretch, and fold dough or similar mixture to make it smooth. During kneading, bread dough becomes elastic, fondant becomes smooth and satiny.

Marinate. To let foods stand in a liquid (usually mixture of oil with vinegar or lemon juice) to add flavor or make more tender.

Pare. To remove skins or peel from fruit or vegetable; peel.

Parboil. To boil until partly cooked.

Poach. To cook gently in liquid at simmering temperature so that the food retains its shape.

Pot-Roast. To cook large pieces of meat by braising.

Pressure Cook. To cook food in water and or steam in a pressure saucepan or canner at temperatures above 212°F (100°C).

Reconstitute. To restore concentrated foods to their original state; for example, to restore frozen concentrated orange juice to liquid form by adding water.

Rehydrate. To soak or cook to make dehydrated foods take up the water they lost during drying.

Roast. To bake in hot air (usually oven) without water to cover.

Sauté. To brown lightly in fat.

Scald. To heat liquid to about 149°F (65°C).

Simmer. To cook in liquid just below the boiling point, at temperature of 185–210°F (85–98°C). Bubbles form slowly and break below the surface.

Steam. To cook food in steam, without pressure. Food is placed on a rack or a perforated pan over boiling water in a covered container.

Stew. To cook, covered at simmering temperature, in a small amount of liquid.

Stir, Blend, Mix. To combine several ingredients to affect an even distribution throughout.

Whip. To beat rapidly to incorporate air.

Appendix L
Buying Guide[1]

VEGETABLES AND FRUITS

A serving of a vegetable is ½ c (120 ml) cooked vegetable unless otherwise noted. A serving of fruit is ½ c (120 ml) fruit; 1 medium apple, banana, peach, or pear; or 2 apricots or plums. A serving of cooked fresh or dried fruit is ½ c (120 ml) fruit and liquid.

Fresh Vegetables	Servings per lb (454 g) As Purchased
Asparagus	3 or 4
Beans, lima[2]	2
Beans, snap	5 or 6
Beets, diced[3]	3 or 4
Broccoli	3 or 4
Brussels sprouts	4 or 5
Cabbage	
Raw, shredded	9 or 10
Cooked	4 or 5
Carrots	
Raw, diced or shredded[3]	5 or 6
Cooked[3]	4
Cauliflower	3
Celery	
Raw, chopped or diced	5 or 6
Cooked	4
Kale[4]	5 or 6
Okra	4 or 5
Onions, cooked	3 or 4
Parsnips[3]	4
Peas[2]	2
Potatoes	4
Spinach[5]	4
Squash, summer	3 or 4
Squash, winter	2 or 3
Sweet potatoes	3 or 4
Tomatoes, raw, diced or sliced	4

[1] Adapted from Buying Food, Home Economics Research Report No. 42 USDA, 1978.
[2] Bought in pod.
[3] Bought without tops.
[4] Bought untrimmed.
[5] Bought prepackaged.

Canned vegetables	Servings per Can 1 lb (454 g)
Most vegetables	3 or 4
Greens, such as kale or spinach	2 or 3

Frozen Vegetables	Servings per Package 9 or 10 oz (252–280 g)
Asparagus	2 or 3
Beans, lima	3 or 4
Beans, snap	3 or 4
Broccoli	3
Brussels sprouts	3
Cauliflower	3
Corn, whole kernel	3
Kale	2 or 3
Peas	3
Spinach	2 or 3

Dry Vegetables	Servings per lb (454 g)
Dry beans	11
Dry peas, lentils	10 or 11

Fresh Fruit	Servings per Market Unit, A.P.
Apples, Bananas, Peaches, Pears and Plums	3 or 4/lb (454 g)
Apricots; Cherries, sweet; Grapes, seedless	5 or 6/lb (454 g)
Blueberries	4 or 5/pt (480 ml)
Raspberries, Strawberries	8 or 9/qt (950 ml)

Frozen Fruit	Servings/Package 10–12 oz (300–360 ml)
Blueberries	3 or 4
Peaches	2 or 3
Raspberries	2 or 3
Strawberries	2 or 3

Canned Fruit	Servings/Can 1 lb (454 g)
Served with liquid	4
Drained	2 or 3

Dried Fruit	Servings/Package 8 oz (224 g)
Apples	8
Apricots	6
Mixed Fruits	6
Peaches	7
Pears	4
Prunes	4 or 5

Meat Products	Size of Serving	
Beef, ground	4 servings/lb	(454 g)
Chicken breast halves	2.75 halves/lb	(454 g)
	1 half = 3 oz	(85 g)
Chicken thighs	4.5 thighs/lb	(454 g)
(cooked meat)	2 thighs = 3 oz	(85 g)

Cereals and Cereal Products	Size of Serving		Servings per lb (454 g)
Flaked corn cereals	1 c	240 ml	16
Other flaked cereals	¾ c	180 ml	21
Puffed cereals	1 c	240 ml	32–38
Cornmeal	½ c	120 ml	22
Wheat cereals			
Coarse	½ c	120 ml	16
Fine	½ c	120 ml	20–27
Oatmeal	½ c	120 ml	16
Hominy grits	½ c	120 ml	20
Macaroni and noodles	½ c	120 ml	17
Rice	½ c	120 ml	16
Spaghetti	½ c	120 ml	18
Flour (all purpose)	1 c	240 ml	4
Cornmeal, dry	1 c	240 ml	3

Miscellaneous	cups/lb 240 ml/454 g
Hydrogenated fats	2½
Butter, margarine, lard	2
Oil	2
Cottage cheese	2
Granulated sugar	2
Brown sugar, packed	2⅓
Confectionery sugar	3½
NDM: 1⅓ c powder + 3¾ c water = 1 qt (320 ml + 900 ml) = 960 ml fluid	

Appendix M
Fruit and Vegetable Availability

Legend:
- ☐ Means supplies are scarce or nonexistent.
- ▨ Means supplies are moderate.
- ▦ Means supplies are plentiful.
- ■ Means supplies are exceptionally abundant.

(Codes: S = scarce, M = moderate, P = plentiful, A = abundant)

COMMODITY	Jan	Feb	Mar	Apr	May	June	July	Aug	Sept	Oct	Nov	Dec
APPLES	P	P	P	P	M	M	M	M	P	P	P	P
APRICOTS	S	S	S	S	S	A	M	S	S	S	S	S
ARTICHOKES	M	M	P	P	P	M	M	M	M	M	M	M
ASPARAGUS	S	M	A	A	P	M	S	S	S	S	S	S
AVOCADOS	P	P	P	P	P	P	P	P	P	P	P	P
BANANAS	P	P	P	P	P	P	P	P	P	P	P	P
BEANS, SNAP	M	M	M	M	P	P	P	M	M	M	M	M
BEETS	M	M	M	M	P	P	P	P	P	P	M	M
BERRIES, MISC.*	S	S	S	S	M	A	M	S	S	S	S	S
BLUEBERRIES	S	S	S	S	S	P	A	M	S	S	S	S
BROCCOLI	P	P	P	P	M	M	M	M	P	P	P	P
BRUSSELS SPROUTS	P	M	M	S	S	S	S	S	M	P	P	P
CABBAGE	P	P	P	P	P	P	P	P	P	P	P	P
CANTALOUPES	S	S	S	S	M	P	A	A	P	M	S	S
CARROTS	P	P	P	P	P	P	P	P	P	P	P	P
CAULIFLOWER	M	M	M	M	M	M	M	M	M	P	P	M
CELERY	P	P	P	P	P	P	P	P	P	P	P	P
CHERRIES	S	S	S	S	M	A	P	M	S	S	S	S
CHINESE CABBAGE	M	M	M	M	M	M	M	M	M	M	M	M
CORN, SWEET	S	S	S	M	M	P	P	P	P	M	S	S
CRANBERRIES	M	S	S	S	S	S	S	S	S	P	A	P
CUCUMBERS	P	P	P	P	P	P	P	P	P	P	P	P
EGGPLANT	M	M	M	M	M	M	M	M	M	M	M	M
ESCAROLE-ENDIVE	P	P	P	P	P	P	P	P	P	P	P	P
ENDIVE, BELGIAN	M	M	M	M	M	S	S	S	M	M	M	M
GRAPEFRUIT	P	P	P	P	P	M	M	M	M	M	P	P
GRAPES	S	S	S	S	S	M	M	P	A	A	A	M

*Mostly blackberries, dewberries, raspberries.

Charts on pp. 274–275 courtesy of USDA.

continued

Appendix M (cont'd)
Fruit and Vegetable Availability

COMMODITY	Jan	Feb	Mar	Apr	May	June	July	Aug	Sept	Oct	Nov	Dec
GREENS												
HONEYDEWS												
LEMONS												
LETTUCE												
LIMES												
MUSHROOMS												
NECTARINES												
OKRA												
ONIONS, DRY												
ONIONS, GREEN												
ORANGES												
PARSLEY & HERBS**												
PARSNIPS												
PEACHES												
PEARS												
PEPPERS, SWEET												
PINEAPPLES												
PLUMS-PRUNES												
POTATOES												
RADISHES												
RHUBARB												
SPINACH												
SQUASH												
STRAWBERRIES												
SWEETPOTATOES												
TANGERINES												
TOMATOES												
TURNIPS-RUTABAGAS												
WATERMELONS												

**Includes also parsley root, anise, basil, chives, dill, horseradish, others.

Information courtesy of United Fresh Fruit and Vegetable Association, Washington, D.C.

Appendix N
Spice and Herb Chart

Herbs & Spices	Main Dish	Salads	Sauces	Vegetables
Spices				
Allspice	Beef pot roast, duck, turkey or chicken, fish	Fruit Salad	Tomato	Beets
Cayenne	Beef, stews, chicken, seafood	All varieties except fruit	Meat, vegetable	All vegetables
Chili powder	Beef, hamburgers, meatloaf, chili con carne	Bean salad	Mexican type	Corn
Cloves	Pork, ham, boiled beef, pot roast		Tomato, sweet/sour	
Curry	Meat, fish, poultry, lamb, veal, fish or shrimp chowders	Chicken salad	Vegetable	Rice, creamed onions
Ginger	Pork, chicken	Fruit salad	Dessert	Squash
Mace	Poultry stuffing, veal	Fruit salad	Fish, poultry, veal	Potatoes
Dry mustard	Beef, hamburgers, chicken, tuna, egg	Chicken, egg, tuna, macaroni, potato	Fish, vegetable	Cabbage
Nutmeg	Chicken stew, beef stew, creamed dishes	Fruit salad	Dessert, Fruit sauces; pudding	All vegetables except cabbage family
Paprika	Meat, fish, poultry, veal, creamed dishes	All except fruit salad	All gravies and sauces	All vegetables
Pepper	Meat, fish, poultry, veal	All except fruit salad	All gravies and sauces	All vegetables
Herbs				
Basil	Tomato, egg, fish, chicken cacciatore, beef stew	Vegetable salads with marinades	Tomato	Cucumbers, green beans, zucchini
Chives	Creamed dishes, fish, eggs	Potato salad, green salad	Creamed type	Potatoes
Dill	Fish	Potato, vegetable	Creamed type	Green beans, cucumbers, cabbage, carrots
Marjoram	Italian type, tomato, beef, lamb, fish	Salad dressings	Tomato, brown	Broccoli, green beans, peas, eggplant
Oregano	Italian type, tomato, meatloaf, pork, veal, pot roast	Vegetable salads, marinades, bean salad, salad	Tomato, fish	Tomato, broccoli, zucchini, eggplant
Parsley	All	All except fruit salad	All except fruit	All
Rosemary	Beef, pork, fish, lamb		Vegetable, meat and fish gravies	Cauliflower, potatoes, turnips
Sage	Pork, poultry, goulash beef stew	Vegetable salads with marinades	Meat, chicken, pork gravies	Mushroom, broccoli, cabbage, onions, cauliflower
Thyme	Beef, pork, chicken, fish, beef stews, fish chowders, fish soups	Vegetable salads with marinades, salad dressings	Brown	Creamed onions
Tarragon	Eggs, poultry, fish	Salad dressings	Creamed type	Potatoes

Courtesy of Campbell Soup Company

Appendix O
Plant Proteins

Meat, fish, eggs, and milk occupy a prominent place in the American diet. Although these foods are excellent sources of protein and several other nutrients, *moderation* of their consumption is advocated for several reasons. The recent U.S. Dietary Guidelines suggest moderation of intakes of total fat, saturated fat, and cholesterol. These animal protein sources contribute a significant portion of these substances to our diets. Further, production of animal protein is expensive. About 10 pounds of grain are required to produce one pound of meat. Accordingly, for the consumer, the price of animal foods is high. Shrinking food resources and rising food prices oblige us to question extensive use of animal foods to meet protein needs. Less expensive plant proteins can adequately meet protein needs if nutritional principles are clearly understood.

Recall that protein per se is not an essential nutrient. Rather, dietary protein provides amino acids and nitrogen. Thus, the protein value of a specific food depends on the total quantity of protein present and the quality of the amino acid pattern. At least 22 different amino acids are used in body processes. Some of these amino acids are synthesized by body cells from commonly available materials. Nine are called essential amino acids because the body cannot synthesize them from any materials. The nine essential amino acids are valine, lysine, threonine, leucine, isoleucine, tryptophan, phenylalanine, histidine, and the sulfur containing amino acid, methionine. These nine must be obtained from food *performed, ready-to-use,* and *in appropriate amounts.* Further, for efficient utilization, the body requires all nine essential amino acids to be ingested at about the same time. *Protein quality* of a single food is judged by its capacity to provide the essential amino acids in appropriate amounts.

Single foods differ in protein quality. The nutritive excellence of such foods as meat and milk is due to their complete amino acid pattern. For this reason, animal foods are said to have *complete proteins.* Gelatin is an exception to the usual high quality of animal foods; it is an *incomplete protein* because one essential amino acid is missing.

Many plant foods have substantial amounts of protein. Although the nine essential amino acids are present, some are not present in sufficient amounts. Thus, plant proteins are classed as *partially complete.* The most frequently limiting amino acids are lysine, threonine, tryptophan, and sulfur-containing amino acids. Individual foods differ as to their essential amino acid composition. These differences are summarized in Table O.1. Since threonine is generally adequate if needs for the other three are met, it is not included in the summary.

To meet body needs, it does not matter whether the essential amino acids come from a single food or from a combination of foods. The key to using plant proteins to meet protein needs is understanding their amino acids composition (see Table O.1). A few plant proteins, such as soy, have good amounts of the essential amino acids. Thus, by increasing the quantity of plant proteins like soy eaten at one time, essential amino acid needs can be met. Plant proteins can be combined with a small amount of animal protein or two or more plant proteins can be combined so there is *mutual supplementation* of the amino acids of the different foods. When two foods supplement each other in amino acid patterns, the combination gives greater protein quality than either could provide if eaten alone. This concept is illustrated in Figure O.1.

Because animal foods generally have a complete amino acid pattern, they complement most plant foods. Dairy foods are a particularly effective complement of plant protein because they are good sources of

TABLE O.1

Food Groups	Amino Acids	
	Good Source of	Poor Source of
LEGUMES	Lysine	Tryptophan S-C[b]
Soybeans	Lysine[a] Tryptophan[a]	S-C
Dry beans	Lysine[a]	Tryptophan S-C
NUTS–SEEDS	Tryptophan S-C	Lysine
Peanuts	Tryptophan	Lysine S-C
Sesame seed	Tryptophan S-C[a]	Lysine
CEREALS–GRAINS	Tryptophan S-C	Lysine
Cornmeal	S-C	Lysine Tryptophan
Whole wheat flour	Tryptophan S-C	Lysine
Wheat germ	Lysine[a]	Tryptophan S-C
Rice	Tryptophan S-C	Lysine
EGGS	Lysine[a] Tryptophan[a] S-C[a]	
MEAT, FISH, POULTRY	Lysine[a] Tryptophan S-C	
DAIRY	Lysine[a] Tryptophan[a] S-C[a]	

[a] Superior.
[b] S-C sulfur-containing amino acids.

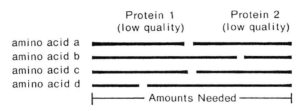

FIGURE O.1 Combining protein 1 & 2 provides good-quality protein.

lysine. A small amount of meat improves the protein quality of grains and legumes. Cereal and milk, macaroni and cheese, spaghetti and meatballs, peanut butter, and milk are examples of complementing plant foods with a small amount of animal food.

Combining two or more plant proteins to obtain high-quality protein depends on matching the amino acid strengths and weakness of individual foods. Nuts, seeds, and grains are generally low in lysine and relatively good in tryptophan and sulfur-containing amino acids. In general, legumes are good sources of lysine and poor sources of tryptophan and sulfur-containing amino acids. Exceptions to those generalizations are noted in Table O.1. With complementary combinations of plant foods, like rice and beans, the resulting amino acid pattern can be equal to animal foods. However, evaluation of total nutrient intake is important when plant foods are extensively substituted for animal foods. Particular attention should be given to amounts of vitamin D, B_{12}, riboflavin, and minerals provided in an all vegetarian diet. Table O.1 provides a starting point for investigating various protein food combinations. More detailed information is given in references cited below.

Use of plant foods as major protein sources is not a new concept. The traditional dishes of many cultures illustrate the concept's extensive application over the years. Currently, the Dietary Guidelines, the world food shortages, and high food prices re-emphasize the value of plant proteins. Creative use of the principles of mutual supplementation not only spares animal foods and stretches the food dollar without sacrifice of protein quality but also opens up a range of new eating experiences.

REFERENCES

Guthrie HA. *Introductory Nutrition,* 7th ed. St. Louis, MO: Times Mirror/Mosby College Publishing, 1989.

Lappé FM. *Diet for a Small Planet,* 20th Ed. New York: Ballantine Books, 1991.

Robertson L, Flinders C, Ruppenthal B. *The New Laurel's Kitchen.* Berkeley, CA: Ten Speed Press, 1986.

Whitney EN, Rolfes SR. *Understanding Nutrition,* 7th ed. St. Paul, MN: West Publishing Co., 1996.